Informational Peripheries

Informational Peripheries

Rethinking the urban in a digital age

Edited by Ayona Datta and
Fenna Imara Hoefsloot

UCLPRESS

First published in 2025 by
UCL Press
University College London
Gower Street
London WC1E 6BT

Available to download free: www.uclpress.co.uk

A CIP catalogue record for this book is available from The British Library.

ISBN: 978-1-80008-887-0 (Hbk)
ISBN: 978-1-80008-888-7 (Pbk)
ISBN: 978-1-80008-889-4 (PDF)
ISBN: 978-1-80008-890-0 (epub)
DOI: https://doi.org/10.14324/111.9781800088894

Printed and bound by CPI Group (UK) Ltd, Croydon, CR0 4YY

Contents

Theme II: territorialising the informational periphery

Theme III: informational agency from the peripheries

List of figures

List of tables

List of contributors

Julie-Anne Boudreau is a Senior Researcher at the Institute of Geography, National Autonomous University of Mexico. She is co-director of PATIO lab, a space for participatory research, and director of TRYSPACES, a transdisciplinary team of 60 researchers, students, multimedia artists, social workers, urban planners and young people from Mexico City, Montreal, Paris and Hanoi. She was appointed a CIFAR Fellow of the 'Humanity's Urban Future' research programme in 2023. She is a recipient of the Sor Juana Inés de la Cruz Award (2023).

Calvin King Lam Chung is an Assistant Professor in the Department of Geography and Resource Management at The Chinese University of Hong Kong. His research examines the urban politics of the environment, with a geographical focus on China and a sectoral focus on planning, infrastructure and industries. His ongoing projects unpack the politics of environmental data and municipal green bond issuance, respectively, in Chinese cities.

Thomas Cowan is an Associate Professor of Urban Geography at the University of Nottingham, UK, and the author of *Subaltern Frontiers: Agrarian city-making in Gurgaon* (Cambridge University Press, 2022). His research has focused on agrarian urbanism, property-making, dispossession and digital infrastructure in postcolonial India. He is the Principal Investigator of the British Academy Wolfson-funded project 'Digital Enclosures: Automating property in India'. His writings have appeared in *Antipode, City, International Journal of Urban and Regional Research, Urban Geography, Gender, Place and Culture* and *Economic & Political Weekly*.

Jingya Dai is a PhD student in the Department of Community, Culture and Global Studies at the University of British Columbia Okanagan, Canada. Her areas of interest include political ecology, environmental governance, and urban and digital geographies. Her research centres on examining China's environmental governance from an institutional perspective and the implications of digital innovations on state–economy–environment relations.

Ayona Datta, FAcSS is Professor in Human Geography at UCL, UK. She researches the intersections of urban futures, gender citizenship and smart cities, focusing on new configurations of gender–power relations in the digital-territorial margins. She is author of *The Illegal City* (Ashgate Press, 2012), co-editor (with Ola Söderström) of *Data Power in Action* (Bristol University Press, 2023), co-editor (with Abdul Shaban) of *Mega-urbanisation in the Global South* (Routledge, 2017) and co-editor (with Katherine Brickell) of *Translocal Geographies* (Ashgate Press, 2011). Her current ERC Advanced Grant on 'Regional Futures' seeks to advance theoretical and empirical insights into the relationship between digitalisation and urbanisation in the Global South. She received the Busk Medal from the Royal Geographical Society (with IBG) in 2019 for her contributions to the understanding of smart cities.

Juan Angel Demerutis Arenas is a Professor at the University of Guadalajara, Mexico. His research interests include public policies for sustainable urbanism, and urban and territorial planning. From 2012 to 2015 he served as Director of Planning and Territorial Planning in the municipality of Zapopan, where he implemented innovative instruments for sustainable urban development. He has been a member of the Mexican National System of Researchers since 2014. He is also a member of various collaborative research communities, including the metropolitan governance network CONACYT and the National Laboratory of Housing and Sustainable Communities.

Jesus Flores is a Postdoctoral Research Fellow at the University of Guadalajara, Mexico, working within the ERC-funded 'Regional Futures' project. His main area of research focuses on the social construction of risk and how socio-spatial dynamics produce unsafe conditions for vulnerable people in the urban context. He has more than nine years' experience of researching the Guadalajara Metropolitan Area. In addition, he has worked as an external consultant for public agencies and coordinated several state, regional and municipal planning instruments.

Catherine Gateri is a researcher, land valuer and urban and regional planner. As a researcher, she has participated in numerous research activities in land administration, land management, urban and regional planning, transport and governance. She has taught at the Technical University of Kenya, Kigali Institute of Science and Technology, Makerere University and Kenyatta University. Working as a physical planner in Kenya's Ministry of Lands from 2003 to 2007 and in private practice, she has been involved in the preparation of spatial development plans in the public domain.

D. Asher Ghertner is Professor in the Department of Geography at Rutgers University, USA. He is the author of *Rule by Aesthetics: World-class city making in Delhi* (Oxford University Press, 2015), co-editor (with Daniel Goldstein and Hudson McFann) of *Futureproof: Security aesthetics and the management of life* (Duke University Press, 2020) and co-editor (with Robert Lake) of *Land Fictions: The commodification of land in city and country* (Cornell University Press, 2021). He has conducted long-term fieldwork and legal-archival research on urban informality, property making and crisis ecologies in postcolonial India. He was a 2023–4 Fulbright-Nehru Scholar based in Goa, India, studying digital enclosures and the legal geography of land speculation.

Shoshana Goldstein is a Assistant Professor of Urban Studies at Trinity College, Hartford, USA. Her work explores environmental and mobility justice for interstate migrants in urban India. She previously taught at the Princeton University School of Architecture as a Mellon Fellow and affiliate with the M. S. Chadha Center for Global India. She holds a PhD in City and Regional Planning from Cornell University, USA.

Fenna Imara Hoefsloot is a Research Fellow at UCL, UK. Her research focuses on the relationships between the digitalising state, knowledge infrastructures and territorial politics. She has researched how digital infrastructures reproduce or restructure inequalities in urban water and land governance and urban planning. In exploring how digital technologies and planning support systems can be reshaped to inform a city that meets everyone's needs, she collaborates with residents and civil society organisations to design new technologies and pilot interventions in digital and infrastructural space.

Dennis Mbugua Muthama is a Postdoctoral Research Fellow in the ERC-funded 'Regional Futures' project. He is an academic, a practising land administration professional and a researcher. His research interests include land administration, urban governance and hybrid land governance. His PhD explored how land tenure administration is undertaken in peri-urban Mombasa, Kenya. As an academic, he has taught at the University of Nairobi, Kenya. He is a Full Member of the Institution of Surveyors of Kenya and the Chartered Institute of Arbitrators.

Julien Migozzi is an economic and urban geographer and an Urban Studies Foundation Postdoctoral Research Fellow at the University of Oxford, UK, and a Visiting Scholar at the Department of Sociology of UC Berkeley, USA. His research examines the urban process under digital capitalism, questioning how housing and financial markets reshape cities, markets and inequalities, with a particular emphasis on the stratifying effects of digital technologies in South Africa. He is a co-author of the *Atlas of Finance* (Yale University Press, 2024).

Josie Wittmer is an Assistant Professor at Memorial University of Newfoundland, Grenfell Campus, Canada. She is a feminist urban geographer working on issues of urban sustainability, labour precarity and environmental health. Her ongoing projects focus on urban infrastructural transformations, digital sustainability solutions, and solid waste and wastewater governance in Canada and South Asia.

Jiang Xu is a Professor in the Department of Geography and Resource Management at The Chinese University of Hong Kong. Her main research area is critical urban and regional studies, with a focus on China's inland region and the Greater Bay Area. Her recent works combine the political economy tradition and a poststructuralist approach as a novel perspective to unravel how emerging industries (such as the big data sector) have (re)shaped China's urban and regional transformation.

Acknowledgements

The idea for this edited volume came about in part from a scoping paper on the 'digitalising state' written by Ayona Datta in the *Progress in Human Geography* journal in 2022, and in part during a co-organised double panel session on 'Regional Futures' at the American Association of Geographers (AAG) annual meeting in March 2023 in Denver, Colorado. We are grateful to the presenters of the sessions and for the stimulating conversations that took place during those sessions, including comments from the discussant, D. Asher Ghertner, who has written the Epilogue to this volume. During AAG 2023, Ayona Datta was also invited to deliver the *Asian Geographer* journal plenary entitled 'Informational Periphery', which gave the name of this volume as well as the opportunity to develop the talk into a chapter in this volume. The editors also co-organised a double session entitled 'Informational Peripheries' at the Royal Geographical Society (RGS-IBG) annual conference in August 2023, where a number of the chapters in this volume took shape. We gratefully acknowledge the foundational contributions to this book by all those who presented during those sessions.

The seeds of the volume were planted during the 2021–3 British Academy-funded Tackling Global Challenges project (ref: TGC\200118) 'Digitising the Periphery: Co-producing a toolkit for digital democracy and inclusive urbanisation' (Ayona Datta (Principal Investigator), Abdul Shaban (Co-Investigator), Sheema Fatima (Postdoctoral Research Fellow)). The development of the concept of the informational periphery was made possible through generous funding provided by the European Research Council (ERC) Advanced Grant awarded to Ayona Datta for the project 'Regional Futures: The territorial politics of digitalisation-as-urbanisation in the Global South' (ID: 101019318), running from 2022 to 2026. This funding has provided the time, resources and empirical material to conceptualise and produce this volume. The 'Digitising Peripheries' symposium in May 2023, funded by the British Academy project, and the 'Digitalising State' symposium in September 2023, funded by the ERC project, offered valuable opportunities for us to present and receive feedback on chapters of this book.

The ideas developed in the book on the intersections of geography, information and the state benefited greatly from our collaborations and discussions with the Regional Futures team, some of whom have also contributed chapters in this volume: Raul Agraz, Juan Demerutis, Jesus Flores, Catherine Gateri, Neha Gupta, Keya Kunte, Dennis Mbugua Muthama, Mariana Reyes Carranza, Abdul Shaban and Melissa Wanjiru. We also thank Gabriela Bazzo, our project administrator, for supporting all the project activities, including the symposium at UCL in 2023. Alongside team members, we thank the Advisory Board of the Regional Futures project – Julie-Anne Boudreau, Rob Kitchin, Ursula Rao, Jennifer Robinson and Pontus Westenberg – for their insightful and constructive comments during our meetings and beyond.

Some fortuitous funding from UCL Global Engagement Funds in June 2024 enabled us to organise a knowledge exchange visit between Ayona Datta at UCL and Professor Ursula Rao at Max Planck Institute for Social Anthropology in Halle. A Visiting Fellowship at UCL undertaken by Professor Wen-I Lin of the National Taipei University facilitated several conversations on this volume and beyond. Special thanks therefore go to Ursula Rao and Wen-I Lin for their generous feedback on the introductory chapter during a workshop funded by the Global Engagement Funds at UCL. Their comments contributed immensely to developing the contribution of the concept of informational periphery to the wider discussions on digitalisation and urbanisation.

This book would not have been possible without the support, cooperation and assistance of several collaborators, co-conspirators and colleagues, for which we are eternally grateful. Both editors are grateful for several opportunities to present ideas from this volume in full or in part in several academic venues. Fenna Hoefsloot is grateful for the opportunity to present her work at the UCL Human Geography Research seminar, and for the generous feedback and suggestions provided by UCL colleagues. Ayona Datta appreciates the feedback from audiences at several engagements, including at the inaugural lectures for 'Exeter Urban' (University of Exeter) and the City Collaboratory launch (University of Lausanne), and keynote engagements at the 2023 UTA-Do Conference in Nairobi, the 2023 Regional Studies Association Winter Conference in London, the 2023 Annual Lecture of the British Institute of Eastern Africa in London, and the 2024 'Urban Tech for Spatial Equity' UNITAC/UN-Habitat Conference in Hamburg. We are grateful for these invitations, and for the feedback we received, which has helped us develop our ideas in this volume.

Finally, we greatly appreciate the support and encouragement of the commissioning editors at UCL Press. We thank Elliot Beck, Chris Penfold and Robert Davies for their guidance in bringing this book to press, our exemplary copy-editor Ruth Massey, and the anonymous reviewers of the book proposal and draft manuscript.

1
An introduction to informational peripheries

Ayona Datta and Fenna Imara Hoefsloot

On 15 December 2023, citizens in the small textile town of Bhiwandi, located on the northern outskirts of Mumbai, gathered outside the Municipal Corporation building to demand the revocation of the Development Plan (DP) 2023–43. This plan, which had been developed following two years of consultation and assessment by private consultants, was still awaiting approval by the Bhiwandi Municipal Corporation. However, the process leading to its creation had been shrouded in secrecy – citizens, community leaders and local professionals had no opportunity to observe or provide feedback on the plan. Despite this, the municipality claimed to have conducted a comprehensive consultation process that resulted in the zoning and criteria for Bhiwandi's future development.

The DP 2023–43 was a significant milestone in Bhiwandi's history and a source of pride for the municipality, which for several decades had been bypassed by Mumbai's regional development plans. The corporation had struggled to cope with rapid 'unplanned' urbanisation in and around the town and often attributed this to its inability to 'see' territorial expansion using simple analogue survey methods. For the first time, the Bhiwandi Municipal Corporation had produced a DP (albeit with the help of private consultants) using relatively sophisticated digital technologies such as drones, remote sensing and geographic information system (GIS) mapping. These new spatial planning technologies were a watershed for a relatively small corporation that had been plagued by decades of mismanagement, resource shortages and limited capacity. So why were the citizens agitating?

The month-long protests and sit-ins by citizen groups in Bhiwandi were directed at the informational gaps and inaccuracies in the DP that directly affected their lives and livelihoods. Their grievances were related to the use of digital technologies, which they claimed had misrepresented existing land use in Bhiwandi. Private consultants recruited by the municipality had used drones and camera-equipped cars to map the area, but they had been unable to enter the narrow alleys and lanes, leaving parts of the city unmapped and invisible in the DP. Further, they had relied on the outdated 2003 DP as the foundation for the new plan, despite the fact that Bhiwandi's rapid urbanisation had not followed any of the zoning and development stipulations set out in earlier plans. Citizens demanded a return to physical survey methods of mapping existing land use in Bhiwandi and a process of citizen consultation to plan Bhiwandi's future.

The events in Bhiwandi illustrate the ways in which control over informational flows determines how territories are seen, represented and governed. Who has access to information, who gets to direct informational flows and who determines what sorts of information will flow – these are forms of power that are crucial to determining how urban futures will unfold. Informational power is arguably more significant than other forms of social, infrastructural or territorial power since it can be exerted simultaneously across physical, social and digital realms. Informational power is infrastructural, social, physical, representational and territorial. Those without informational power are peripheralised – they are denied agency over the urban transformations unfolding around them.

In this book we refer to the diverse processes of peripheralisation evident in a digital age with the term 'informational peripheries'. Following Datta (2023; 2024b), we define 'informational peripheries' as a combination of digital, material, territorial, infrastructural and social marginalisations that emerge from informational control, redlining, manipulation and blockading. We argue that in a digital age, the periphery is more than just a geographical location. Seen through an informational lens, it emerges as a temporal construct that continually evolves across virtual and analogue spaces. Informational peripheries are determined by a lack of access to informational flows, as well as a relational positionality of expulsion from crucial informational exchanges across geographic space-times. They emerge from the residual spaces of flows, circuits and networks of information and the laws, policies and governance systems that channel information away from particular geographies. The informational periphery is fragmented, diffuse and dispersed and cannot be fixed or determined through simple cause-and-effect analysis.

It is entangled with regimes of capital, land transformations and infrastructural disconnections across several scales. The informational periphery, then, is a heuristic that stands for the digital-urban condition today. Because it cannot be fixed in location or time, informational peripheries are multitudinous, diverse and continually shape-shifting.

The agitations of Bhiwandi's citizens remind us that across the world, peripheral spaces and communities are becoming the experimental testbeds of new smart technologies for visualisation, mapping and urban governance. The rapid rise of sophisticated data-based technologies in public services in recent decades has been accompanied by skyrocketing investments in data-driven administrative reforms and an incredible growth in companies specialising in facilitating these services, including software developers, system architects and geo-information consultants. These reforms and investments are rationalised by calls to speed up service provision, achieve greater efficiency and gain deeper knowledge to anticipate future challenges (Coletta and Kitchin 2017). But the reach of these tools and their impacts often coincide with increased surveillance of the poor, the curtailment of opportunities to audit policies (Eubanks 2018) and exploitation of people on the social-economic-geographic margins. This is not a coincidence, and the technologies for urban governance are not neutral. They emerge out of techno-capitalist projects and shape the politics and experiences of being and becoming 'urban' in a digital age.

This book presents a case for 'informational peripheries' as an analytical lens for understanding the new configurations of urban-digital peripheralisations. While the information revolution infiltrates all processes of governance and planning, informational peripheries become both drivers and outcomes of uneven urbanisation processes. The informational periphery is fraught with tensions between erasure/inscription, inclusion/exclusion, (in)visibility and (dis)connectivity in an era when new digital technologies are seen as the future of planning and governance. In the informational periphery, social exclusions are marked by both geographic and informational distance from the state; it includes subjects that are uncountable as well as territories that are unmappable – digitally, socially and geographically. The informational periphery intersects with urban peripheries but also straddles geographic and informational spaces. It attends to diverse and heterogeneous forms of digitalised urbanisation that are taking shape across metropolitan regions in the Global South.

In this book we also argue that, despite the challenges faced by those denied informational power, peripheralisation nonetheless offers possibilities for producing informational flows from below. For example, during the COVID-19 pandemic, a collective of citizens living in Kibera, Nairobi's largest informal settlement, formed Community Mappers. Initially focused on conducting community needs assessments and gauging the impact of the pandemic on young women, Community Mappers has grown beyond these goals. It is now a partner in generating data for infrastructural management within Kibera and has assisted in classifying satellite images to model deprived neighbourhoods across the city. The data collected by Community Mappers literally and figuratively puts Kibera on the map, drawing the attention of the state and supporting its claims to welfare provisions. In the case of Bhiwandi, the absence of spatial data leads to the exclusion of citizens from its urban futures represented on a map; in the case of Kibera, networked connectivity brings a marginal group within a geographic periphery closer to the state.

Informational peripheries, then, are paradoxical spaces that are at once marked by invisibility, exclusion and marginalisation while at the same time being targeted for incorporation into the territory of the metropolitan city. They are materially and epistemologically located in the intersections of multiple imaginaries of the urban, the desire for governability and expansion. The periphery and processes of peripheralisation are not points on the map or linear transformations but anchors of the urban condition at different moments in time. From this perspective, urban spaces and urbanisation are shaped by global political economies, as well as by the small but continuous actions of people, as citizens, as engineers, or through the routine tasks of bureaucrats at work. This urges us to take a deeper look at the practices of accommodation and subversion in the making of the periphery. Data-based systems are ubiquitous in all these spaces, especially in the offices responsible for land, property and resource administration, low-wage and gig workplaces, and the management of public services and infrastructures. This requires thinking about urban transformation not as a singular or linear event but as a temporal and irregular process, oscillating in pace and progress.

This book is a study of this fluid notion of informational peripheries emerging at the nodes of urbanisation, digitalisation and networked societies. As compared to a more topical subject – data – we suggest that informational peripheries are more than data marginalisations. Rather, we approach information as data that has been organised – to which a certain logic has been applied to carry meaning and serve a purpose. In the current digitalising moment, the periphery is produced not just by data flows and blockades, but also by how information is leveraged towards political ends. Collectives of government, local consultants and global companies make maps of new regions for development based on data collected by digital technologies. Digital platforms and integrated databases store and analyse personal data, labour information and register records to control movement, calculate welfare and tax lands, often with poor safeguards regarding privacy and data security (Masiero and Das 2019; Hoefsloot and Gateri 2024). These systems are being integrated into peripheral spaces at an astounding pace, with little political discussion about their impacts and who they serve. We suggest here that the periphery in a digital age needs to be understood through the lens of information – encompassing both its politics and its flows through people, spaces and technologies. This 'urbanization of information' (Shaw and Graham 2017) means that informational power is simultaneously urban, social, political and technological – producing specific inequalities and exclusions across digital and analogue spaces. It fundamentally alters how we might begin to think about the notion of the periphery itself in a digital world.

Periphery in a 'digital turn'

There has been a renewed focus in urban and regional studies on the concept of 'periphery' and the processes of peripheralisation (Gururani 2024; Hauswedell et al. 2019; Phelps et al. 2023). Stemming from Wallerstein's world systems theory, which aimed to capture the geographical dualisms of the Global North versus the Global South, the urban core versus the periphery, and international divisions of labour within extractive and capitalist globalisation, the concept of the periphery has been translated, redefined, localised and critiqued within urban and development studies, research on knowledge production, sociology, and far beyond (el-Ojeili 2015). In the Global South, conceptualisations of the periphery have largely been driven by debates around capital accumulation and dispossession, often placing the periphery in opposition to the core, which has traditionally and theoretically been seen as the locus of industrialised production and wealth accumulation through the processes of empire and neoliberal capitalism (Petrusewicz 2019). Within urban studies, the rich scholarship on the periphery has presented a number of concepts that capture its characteristics and parameters – desakota, suburban, rurban, edge and so on. Seen as comprising industrial zones, business technology parks, gated communities and slums/informalised settlements, among others, peripheries are often understood as the result of land acquisition, autoconstruction, housing programmes and municipalisation (Caldeira 2016; Erman and Eken 2004; Hernandez and Titheridge 2016; Kohli 2004; Pieterse 2018; Sood 2021).

Moving away from the core/periphery dualisms, Caldeira (2016) defines the peripheries in and of themselves and explains how peripheral urbanisation in the Global South is often characterised by its defiance of official policy, the creation of new claims over land and resources, and the redefinition of the relationships between the state and citizens. These particularities allow for the global processes of digitalisation and urbanisation to create new folds in the fabric of the metropolitan city, each with its singular topography of proximity and distance. Peripheries in this framing have been seen as the outcome of 'rapid and unplanned' urbanisation and urban growth – a geographic metaphor to stand for exclusion from the city – and the process of peripheralisation serves to examine the fragmented ways in which territory, politics and actors align in the making of contemporary urban landscapes (Caldeira 2016; Gururani et al. 2021).

While this continued gaze upon the margins of the city has brought about important insights regarding the flows and relationships of capital and investment that enable new actors and urban spaces to emerge, it has yet to engage with the impacts of digitalisation on the periphery. From the digital sentience proposed by smart city scholars (Gabrys 2014) to the geographical expansion of the city noted as suburbanisation (Keil 2017) and sweeping changes initiated through agrarian transformations (Bhagat 2005), how do we reimagine the notion of periphery today when urban spaces are increasingly mediated through the digital? What then constitutes the periphery in a digital age? How does this unfold in the Global South? We argue that a consideration of the impacts of a digital age complicates a geography of the periphery even as the imbalanced and disconnected infrastructures of a digital age produce a far more fragmented and dispersed nature of the urban as we know it.

The importance of thinking expansively about the periphery is articulated by Simone, who conceives of the periphery through 'shared colonial histories, development strategies, trade circuits, regional integration, investment flows and geographical articulation' (Simone 2010, 10). For Simone, peripheries have a double status – as spaces of both 'insufficiency and incompletion' – and while in common usage 'peripheries' refers to the outskirts of cities, Simone urges us to rethink cities in the Global South as also located within the peripheries of urban theory. He proposes that peripheries are frontier or buffer spaces where more 'direct forms of confrontation among cities, regions and nations take shape' (Simone 2010, 41). Similarly, Schmid and Brenner present the periphery not as a geographic space but as an epistemic viewpoint from which the city can be analysed (Schmid 2018; Brenner and Schmid

2014). Seeing from the periphery, it becomes possible to view the extent and reach of urbanisation over resources, people and land outside the traditional view of the city.

In his recent work, Simone (Simone et al. 2023; Simone 2022, 6) has argued that the periphery is the 'surrounds' or the 'extensions' – 'a relational location' rather than a geographic one. Although he does not directly connect this description with digital infrastructures per se, the periphery, through his lens, can be seen as a continuously shifting terrain of lived experience within the nodes of infrastructure and information lockdowns. This may be extended to a digitally mediated geography where peripheries emerge out of the variations in the dialectic relationship between space, society and technology (Zook et al. 2004). This means that in a digital age, the periphery is what Simone and Castán Broto (2022) have called a 'multiform universe', where space is not fixed but constantly created through engagement, detachment or adaptation. We follow Simone's proposition of extension and trace a relational location of the periphery as a mediated space lying in the intersections of social, infrastructural and digital exclusions from the city.

How do peripheries transform through the capacities afforded by digital infrastructures and technologies? In defining the informational periphery we take our cue from Anita Say Chan (2014), who argues that we need to focus on the nature of fragmented and splintered relationships produced across cities and regions that make peripheries both geographically grounded and spatially dispersed. For Chan, peripheries are spaces made through networked power and embody notions of banishment, dispossession, disempowerment and primitive accumulation. This emerges from a position which holds that infrastructures are networked across physical and virtual worlds and that infrastructural territorialities are being deconstructed and reassembled through digital technologies. This diffusion of boundaries between digital and material also draws upon Graham and Marvin's (2001) work on 'splintering urbanism' which produces 'premium networked spaces' and uneven social geographies. Here, the informational periphery becomes at once a site, an instrument and an outcome of information flows.

As detailed in this volume, whether in the segregated neighbourhoods of Cape Town, the informalised settlements of Lima or the newly constructed e-commerce warehouses in Bhiwandi, the urban peripheries of metropolitan regions are not always infrastructurally disconnected, as is often assumed. While urban studies have offered us a rich scholarship to think about peripheries in the Global North and South, the infiltration of information and communication technologies across the far reaches of metropolitan regions offers renewed thinking about the periphery. In this book, we will explore the vastly complex and contradictory digital transformations that are being layered over decades of geographical and social marginalisations to understand how peripheries appear and transform in this 'digital turn' (Ash et al. 2018; Datta 2018).

Structure of the book

In this introductory chapter, we develop the concept of the informational periphery along three main axes: information infrastructures, territorialisation, and reimagining from the peripheries. The chapters that follow contribute across all three axes, offering a historical and critical analysis of informational peripheries. We propose that the empirical and theoretical development of informational peripheries can provide an important vantage point for interrogating the political and technological relations that are at the heart of the emergence of urban peripheries in three important ways: by analysing their production, their territorialisation, and how they are experienced from below. In this book, we undertake two significant reversals. First, we centre the Global South and analyse the multiple ways in which digital technologies take hold in urban contexts. Second, we move from the core to the periphery to understand how digital technologies shape the dynamics between urban and regional landscapes. These dynamics are mediated between technology, the state and everyday life in the informational peripheries.

The book comprises 11 empirical chapters written by early, mid-career and established international scholars, offering both empirical and theoretical insights into the development of informational peripheries across the Global South. It is organised along three main themes: (1) the production of the informational periphery through the grafting of digital information onto colonial and postcolonial paper-based information systems in uneven and unequal ways; (2) territorialising the informational periphery through the actors that shape material forms of informational space; and (3) informational agency from the peripheries, which examines how information infrastructures are generating new spaces for repair, co-optation and manipulation by citizens and civil society organisations. The book concludes that a shift in the lens from urban to informational peripheries provides a critical vantage point for interrogating the political and technological apparatuses that are reconfiguring the notion of peripheries in the digital age globally.

Following the initial conceptualisation of the informational periphery set out in this chapter, the authors in this volume introduce and draw on a range of concepts and metaphors to develop the idea further. From calculable peripheries (Migozzi, Chapter 8) and informational distancing (Wittmer, Chapter 3) to informational worlds (Boudreau, Chapter 11), the authors adopt and adapt the concept to capture how the informational periphery takes shape in each context. For example, Goldstein (Chapter 10) provides a fascinating example of how the informational periphery is not confined to a fixed geographical location but can migrate with the people who embody it. Through the digitalisation of identification systems, such as Aadhaar in India, it becomes easier for people to 'take the data with them' when they change location, facilitating cyclical migration without the necessity to 'become local'. This, as Goldstein explains, creates an ambiguous situation whereby people's circular migration habits are potentially eased without challenging the structures that keep them on the fringes of citizenship and the city. In Chapter 11, Boudreau defines the informational periphery not in relation to the core but by conceptualising the informational core and informational periphery as separate worlds that are built up of different ontologies, epistemologies and sensory apparatuses. Conceptualised as inhabiting not different parts of the city but different urban worlds, a path towards justice does not mean reducing distance or increasing connection but merging ontologies. Consultants, GIS experts, land valuers and lawyers – as well as gig labourers – are all crucial actors in the production of the information systems used to govern urban territories, a process characterised by private-sector involvement and control creep (Kitchin 2020), a lack of allocated resources, and little in-house capacity to digitise vast amounts of paper documents or develop platforms for data management.

This book, then, provides an entry point for understanding the complexities of informational power in a digital era by exploring the slippery nature of spatial, digital and social peripheralisation in the Global South. Through a variety of methods, the chapters examine the political technologies of informational access and restrictions. It is therefore necessary to follow information flows through infrastructures, offices and decision-making processes (Migozzi, Chapter 8) to conduct ethnographies of the archives and trace how paper-based data is digitised and stored, and how information is made history again (Datta and Muthama, Chapter 2), and to explore creative, feminist and decolonial methods that elicit the disparities between the data from the state and the embodied knowledge of people inhabiting the peripheries (Boudreau, Chapter 11; Wittmer, Chapter 3).

Theme I: producing the informational periphery

The first theme in this volume focuses on the production of informational peripheries through simultaneous exchanges between paper and digital systems of information. Paper works in multiple ways within a digital system, not least because human habits of holding and moving paper in files and folders have largely remained despite extensive e-governance initiatives since the 1990s. Matthew Hull (2012) has aptly referred to the urban government of Islamabad as a 'government of paper', pointing to the material and epistemological challenges posed by documents and the documentation of urban planning. Hull explains how documentation is an essential part of colonial and modern government, specifically creating accountability at a distance, recording transactions, and keeping track of movements and changes within the urban territory. This 'bureaucratic memory' layers over newly learnt skills of using computers, entering data, and reading this to make sense and take decisions. As several chapters in this volume argue, this is not a seamless process and the transfer from paper documents to digital platforms reproduces informational gaps and absences which produce new forms of peripheralisation that are simultaneously geographic, material, social and virtual. With the shift to new methods of recording, registering, storing and searching, digital technologies are transforming how states are seeing their subjects and territories – and, equally importantly, producing new informational peripheries.

The chapters in this section illustrate how the informational periphery is produced in different contexts. The authors argue that the informational periphery is a temporal construct as it is produced through a particular intersection of historical, geographical, infrastructural and social marginalisations. It has different origins and legacies and is, therefore, constantly shifting. This fact makes those subject to informational peripheries harder to identify, in turn making it harder to address their multiple forms of precarity. A crucial theme in this invisibility is the rise of new forms of precarious labour. As Datta and Muthama (Chapter 2) argue, the digitisation of paper in Kenya is only possible through the 'invisible' labour of interns and casual workers who are formally outside state employment, but their work is akin to 'gig labour' within state spaces. This 'peripheralised labour' is a paradox, as they are educated but unemployed Kenyan youth contributing on-demand, 'just-in-time' flexible labour towards the delivery of state aspirations of a digital future. Like Datta and Muthama, Wittmer (Chapter 3) looks at labour to unpack how the informational periphery is produced in India through the unequal distribution of precarity, surveillance and tracking actions, following caste, patronage and capitalist relations. Wittmer demonstrates how the informational periphery is produced by selectively rendering the labour of migrant waste workers in Ahmedabad visible through geo-tagging equipment and smartwatches while they continue to be excluded from the state's databases – databases that would help them gain access to housing and welfare. She shows how the fragmented and selective information infrastructures produced through the partnership between the state and the private sector extract data and neglect people.

If labour is required for the production of digital platforms, it is also in the design of these platforms that informational peripheries are perpetually producing new forms of peripheralisations. As Demerutis, Datta and Flores (Chapter 4) illustrate, informational peripheries are produced by the design of land management platforms in Guadalajara, Mexico. On the Visor Urbano platform, funded by Bloomberg Philanthropies, information on land titles and cadastral data-sharing practices by municipalities were uneven and fragmented across the metropolitan region. This was exacerbated by previous informational gaps around agrarian and commons land, as well as the leverage municipalities held around data sharing with the sub-national state. Similarly, Hoefsloot and Gateri (Chapter 5) note how the informational periphery is produced in Nairobi's land management platform – Ardhisasa – through informational gaps in data on informalised settlements. Hoefsloot and Gateri's analysis shows how the platformisation of the state is a power-laden and politically ambiguous practice, where 'elite stakeholders' are at the core and the organisation and people striving for more communal land and information systems are kept in the margins. By excluding paper records and mapping technologies used to administer tenure relationships in informalised lands from the state's land information platform, the digital infrastructure extends the periphery from geographical to informational space.

Together, these chapters describe a stretching of precarious life beyond a government of paper to new, uncharted networks of peripheralisations located, as Datta (2024a) has noted, in 'distant time' and across informational and geographical spaces. They open up ways of seeing the periphery not just through new arrangements of labour relations with the state, but also in the production of peripheries by design. These processes of producing informational peripheries are each distinct. From peripheralised labour within state digitalisation and land administration systems (Datta and Muthama, Chapter 2; Wittmer, Chapter 3) to the design of land management platforms (Demerutis,

Datta and Flores, Chapter 4; Hoefsloot and Gateri, Chapter 5), the authors detail how digital technologies of the state are produced through networks of state and non-state actors, formal and informal labour, data flows, and chokes across and beyond state spaces.

Theme II: territorialising the informational periphery

The idea of the informational periphery enables us to capture contemporary processes of digitalisation as inextricably linked to the territorialisation of informational space. Following Elden (2010), we see territory as a political-technological tool rather than a bounded entity, and, by extension, a contested space of belonging and ownership across digital-material realms. As Elden notes, land and property are often conflated with territory, but in order to understand how territory extends itself beyond ownership of land, we suggest that territory needs to be examined through the informational capacities offered by new technologies of measuring, drawing and representing land and property ownership. The informational age offers us numerous possibilities to think through processes of territorialisation. This is because information and communication technologies, property technologies, financial technologies, land administration and so on are remodelling planning and governance through new visualisation technologies, transforming areas that were once remote and unseen into spaces that are increasingly visible, mapped, calculable and programmable.

The metaphors of the cloud and the web shop have had significant impact in creating a mental picture of the internet and its information infrastructures as being immaterial, floating in mid-air all around us (Amoore 2018; Wyatt 2021). However, as has been firmly established, the digital is very much material, and the flow of all information is dependent on the availability and accessibility of physical infrastructures, often occupying peripheral land, putting pressure on other physical infrastructures, and often maintained by low-wage labourers. Thus, the real impacts of digitalisation are often seen in the metropolitan peripheries. It is in the peripheries that the human, material and environmental costs of urban digitalisation are most deeply felt.

These dynamics are not unique to this digitalising moment. From Scott's classic work (1999) on the legibility-making schemes deployed by the state to make space, resources and people governable, to decolonial scholarship which highlights the multifaceted calculation of territory to create the conditions for colonisation as well as providing indigenous peoples with tools for liberation, claiming land back and engaging everyday knowledge with the terrain (Bryan 2012), scholarship has long highlighted how territory, information and power are intimately linked (McCall et al. 2021). More recently, scholarship on this theme has increasingly explored how digital technologies are structuring space and everyday life (see, for example, Kitchin and Dodge 2011), examining how the geographies of information infrastructures are, and have always been, uneven and clustered, giving some places and people greater visibility and power while enforcing a return to paper for those who remain in the informational periphery (Sriraman 2018).

Periphery as a concept, therefore, creates an indelible dilemma of digital territorialisation, as it proves to be incredibly slippery and impossible to 'fix' in a territory or geographic location. Information technologies enable the periphery to be networked into the city and prised open for urbanisation, while, simultaneously, the periphery presents the most complex challenges of marginal, precarious and expunged lives. In the age of datafication, various sources of information are layered in the periphery, creating a differentiated view of the city from different vantage points (Hoefsloot, Martinez and Pfeffer 2022). The informational periphery as a complex topography of urban forms is produced from this entanglement of social marginalisations, sedimentation of global capital and infrastructural disconnections.

The chapters in this section traverse topics as diverse as labour, environmental protection, housing and property to show how the informational periphery is tied to the politics of urban expansion and how information is also territorialised through the mediating processes of property, infrastructure and capital. This is illustrated in Datta's account (Chapter 6) of the territorialisation of informational peripheries in Mumbai's metropolitan peripheries, which she argues is a form of 'settler colonialism'. In Bhiwandi, a small textile town in Mumbai's peripheries, digital infrastructures are both far-reaching and unreachable. Datta argues that the profusion of warehouses around Bhiwandi is evidence of its territorialisation by a global logistics sector that is simultaneously sticky and displaceable. The territorialisation of a global informational periphery in Bhiwandi is firmly rooted in a settler colonialist dynamic that is territorial and embodied in the pre-existing caste and religious dynamics of a local merchant class. Territorialisation in the warehouses shapes political-technological relationships with land and labour, which are both relegated to the informational peripheries of a global logistics sector. In other words, the informational periphery is where the rhythms of global capital, information infrastructures and local land politics meet and propel the tensions and contradictions of urban transformation.

Chung, Dai and Xu (Chapter 7) further analyse the territorialisation of the informational periphery through the processes of resource extraction. Investigating Guiyang as a peripheral city in China's urban policies, which, unlike China's coastal cities, has not benefited from favourable tax schemes or large-scale infrastructural investments, Chung et al. illustrate how the informational periphery is territorialised on regional as well as national scales to serve the purposes of 'accumulating through information extraction'. They argue that the informational periphery is 'marginalised by and for information flows' as they explain the growing centrality of data infrastructures in environmental policy and its paradoxical aspirations of delivering environmental sustainability while extracting from and devastating natural resources. The production of the informational periphery in their case is geographically and politically removed from the core, as peripheralisation is also a process of environmental transformation, which opens up land to serve as the terrain of global connectivity.

The final two chapters of this section illustrate how informational peripheries are produced from the layering, or stacking, of territory, property and people in particular geographical temporalities. Migozzi (Chapter 8) describes the layering of data and digital technologies to expand the concept of the informational periphery to the 'calculable periphery' in Cape Town. In this configuration, it is not only data and information about the periphery that shapes its (dis)integration – as was the case in the apartheid state – but the automatised calculations of algorithms that steer property markets to create racially segregated urban landscapes. Similarly, Cowan (Chapter 9) explores the introduction of digital technology for property management (PropTech) in India's urban peripheries, and gives a detailed account of how PropTech aims to integrate properties into the capitalist system. This effort, however, encounters tension, as many land and property ownership structures in India are rooted in formal or informal social institutions. In this context, property is often viewed as a communal or collectively owned asset, which does not align with the frameworks of capitalist land governance.

All the authors note that it is not just the material infrastructures – such as fibreoptic cables, sensors and control rooms – that produce the informational periphery, but also the immaterial actors – people, knowledge and narratives – that shape the sedimentation of informational flows on land and territory. The authors highlight how structured or unstructured, state or corporate, openly available or private, data from all directions are being layered together into a 'computational stack' (Bratton 2016), producing particular configurations of informational peripheries.

Non-state, internet-based actors such as Wikipedia, Open Street Map and Google Maps, and intermediaries such as land brokers, surveyors and knowledge consultants, play an important role in creating uneven temporal and geographical spaces of informational flows and blockages.

Theme III: informational agency from the peripheries

The third and final theme of the book focuses on how informational peripheries are experienced, constructed, or subverted from below. In her important book *Change of State*, Sandra Braman observes that '[t]he informational state knows more and more about individuals, while individuals know less and less about the state' (Braman 2006, 314). This is certainly evident in the Global South, as informal and extra-legal arrangements in bureaucracy often produce an overall illegibility of the state (Das and Poole 2004). Across several countries, a lack of data privacy laws and insufficient checks on data extractivism impede the rights of citizens to freedom of information and data sovereignty. On top of that, the proliferation of information infrastructures and policies has widened the asymmetry between what citizens know and what the state knows, leading to a potential erosion of privacy protection, transparency and accountability, and an increase in surveillance through the many technologies of observation by the state (Braman 2006; Milan and Beraldo 2024). As Agrawal and Kumar argue, data in the form of statistics constitute the bedrock of governance and closely reflect the interests of those in power (Agrawal and Kumar 2020). 'Who counts, what counts and how to count' remain inherently political choices that are often unquestioned and unanswered in the periphery. From this perspective, misinformation, malinformation and disinformation are not merely the result of poor practices in generating and sharing information but techniques deployed to assert or destabilise power by both state and non-state actors (Easterling 2016; Scott 1999).

What does it mean to inhabit the informational periphery, with all its complexities, contradictions and temporalities? We acknowledge that much work has been done regarding the role of information infrastructures in structuring the daily work, movement and consumption of people in the geographical and social peripheries (Guma 2022; Heeks and Shekhar 2019; Hoefsloot, Richter et al. 2022; Masiero and Das 2019). For instance, Kwan (2002) highlights the structuring power of information technologies in shaping the geographies of everyday urban life. She presents the case for empirically researching how the impacts of information technologies intersect with gender, race and class dynamics across spaces and social contexts. Other scholars have shown that as informational flows and pathways have changed in the digital era, citizens themselves have become more vocal and active in developing information technologies and leveraging data to pursue their own political and social goals (D'Ignazio and Klein 2020; Milan and Beraldo 2024). Willett and Lang (2018) conceptualise this as 'peripheral discursive agency', capturing the tension in the periphery as the space which is talked about or talked from. They argue that the identity of the periphery is discursively produced within the larger cultural, economic and knowledge structures around space (Willett and Lang 2018). As Graham (2019) notes, digital technologies are tools: they can be used by both state and citizens to advance their political and economic projects, but with very different results. Here, we are proposing that the technologies producing the informational peripheries are appropriated by different communities in different ways, as political tools, with different outcomes.

The three chapters in this section investigate the results of empirical research in Mexico City, Lima and New Delhi to understand how people inhabiting informational peripheries respond in the face of crisis, infrastructural collapse and rapid datafication. The authors note that traditionally marginalised groups are innovating in the informational periphery, appropriating information infrastructures in novel ways. Both Boudreau (Chapter 11) and Hoefsloot (Chapter 12) illustrate how improvisation and grassroots initiatives provide alternatives to top-down digitalisation that are more democratic, better adjusted to daily life, and better aligned to people's lived experiences in the periphery. Boudreau describes how, during the COVID-19 pandemic, state officials were making decisions based on aggregated data, rapidly constructed models and plotted curves rather than street-level knowledge. Both authors argue that the relationship between digitalisation, political economy and place cannot be captured by a single narrative. As Zook et al. (2004) put it, the digital geographies emerging from technological innovation can be both 'empowering (for the people and places able to construct and consume them) and potentially overpowering as institutional and state forces are able to better harness information with growing personal and spatial specificity' (Zook et al. 2004, 156).

The COVID-19 pandemic plays a central role in two of the chapters, not as the primary research subject but as a lens that brings into focus the disparate informational worlds inhabited by urban citizens and policy makers. Goldstein (Chapter 10) and Boudreau (Chapter 11) use the pandemic as a temporal snapshot to reveal the inner workings of information infrastructures and the barriers they create, impeding people's movement and obstructing the flow of knowledge. Amid the increased reliance on digital communication methods during the lockdowns and the repeated focus on data and grafts to get a grip on what was unfolding around us, the pandemic, as a moment of profound crisis and restructuring, mirrors the moment of breakdown described by Star and Ruhleder (1996) as a puncture in time which reveals the workings of infrastructure and the normative aspirations inscribed into the system. In both cases, the use of information infrastructures and the creation of data during the pandemic revealed the production of 'truths' through inscription.

Goldstein's (Chapter 10) research with cyclical migrant communities working in the textile industries on the peripheries of New Delhi showcases how the act of registration, be it on paper or in digital format, solidifies certain issues or events as established truths that are difficult to change once inscribed. While the digitisation of registers in India and the introduction of biometric identification systems might make it easier to access and transfer this information across locations, they do not solve the problems inherent in the 'paper truths' created by mistakes in the data.

While Goldstein's case illustrates how digitalisation reproduced the flaws and limitations of the numerous paper truths in India, Boudreau's research on Mexico City delves into the pluriversality (Escobar 2018) of the informational worlds created through numerical data and embodied knowledge. The ontological and epistemological differences between these two bodies of information create multiple, distinct peripheries, all with their own knowledges, territories and communities. Nevertheless, as Hoefsloot (Chapter 12) illustrates in her analysis of Lima's informational periphery, Mattern's (2021) metaphor of 'grafting' might be a productive way of thinking about how different forms of intelligence can grow together within the city.

As we read these very different chapters, then, we embark on an exploration of both their differences and commonalities – how we can draw insights from the shared experience of inhabiting informational peripheries across the Global South without diminishing their distinct nature. Each chapter offers a fine-grained, detailed account of the imprint that information technologies have left on history, territory and people. Each notices the multi-sectoral interactions and the material and political stories that emerge from within and through the peripheries, attentive to minute disruptors as well as global economies. The shared experience of these distinct narratives is the experience of braiding together local and global, digital and material, past and future, embodied knowledge and data. Each is a story of a rupture created by techno-capitalism, told from the perspective of those who are not at the steering wheel. The chapters show how informational infrastructures have systematically inscribed themselves onto land, homes and bodies. What emerges from each account is clear: the stories told have roots in the past and are unfolding in the future. They provide recordings for current observations, uncertain as to how or when the processes observed will end.

References

Agrawal, A. and Kumar, V. 2020. *Numbers in India's Periphery: The political economy of government statistics*. Cambridge: Cambridge University Press.

Amoore, L. 2018. Cloud geographies: Computing, data, sovereignty. *Progress in Human Geography* 42(1): 4–24. https://doi.org/10.1177/0309132516662147.

Ash, J., Kitchin, R. and Leszczynski, A. 2018. Digital turn, digital geographies? *Progress in Human Geography* 42(1): 25–43. https://doi.org/10.1177/0309132516664800.

Bhagat, R. B. 2005. Rural-urban classification and municipal governance in India. *Singapore Journal of Tropical Geography* 26(1): 61–73. https://doi.org/10.1111/j.0129-7619.2005.00204.x.

Braman, S. 2006. *Change of State: Information, policy, and power*. Cambridge, MA: MIT Press.

Bratton, B. H. 2016. *The Stack: On software and sovereignty*. Cambridge, MA: MIT Press.

Brenner, N. and Schmid, C. 2014. The 'urban age' in question. *International Journal of Urban and Regional Research* 38(3): 731–55. https://doi.org/10.1111/1468-2427.12115.

Bryan, J. 2012. Rethinking territory: Social justice and neoliberalism in Latin America's territorial turn. *Geography Compass* 6(4): 215–26. https://doi.org/10.1111/j.1749-8198.2012.00480.x.

Caldeira, T. P. 2016. Peripheral urbanization: Autoconstruction, transversal logics, and politics in cities of the Global South. *Environment and Planning D: Society and space* 35(1): 3–20. https://doi.org/10.1177/0263775816658479.

Chan, A. S. 2014. *Networking Peripheries: Technological futures and the myth of digital universalism*. Cambridge, MA: MIT Press.

Coletta, C. and Kitchin, R. 2017. Algorhythmic governance: Regulating the 'heartbeat' of a city using the Internet of Things. *Big Data & Society* 4(2). https://doi.org/10.1177/2053951717742418.

Das, V. and Poole, D. (eds). 2004. *Anthropology in the Margins of the State*. Santa Fe, NM: SAR Press.

Datta, A. 2018. The digital turn in postcolonial urbanism: Smart citizenship in the making of India's 100 smart cities. *Transactions of the Institute of British Geographers* 43(3): 405–19. https://doi.org/10.1111/tran.12225.

Datta, A. 2023. The digitalising state: Governing digitalisation-as-urbanisation in the global south. *Progress in Human Geography* 47(1): 141–59. https://doi.org/10.1177/03091325221141798.

Datta, A. 2024a. Distant time: The future of urbanisation from 'there' and 'then'. *Dialogues in Human Geography*. https://doi.org/10.1177/20438206241253567.

Datta, A. 2024b. The informational periphery: Territory, logistics and people in the margins of a digital age. *Asian Geographer* 41(2): 125–42. https://doi.org/10.1080/10225706.2023.2253233.

D'Ignazio, C. and Klein, L. F. 2020. *Data Feminism*. Cambridge, MA: MIT Press.

Easterling, K. 2016. *Extrastatecraft: The power of infrastructure space*. London: Verso.

Elden, S. 2010. Land, terrain, territory. *Progress in Human Geography* 34(6): 799–817. https://doi.org/10.1177/0309132510362603.

el-Ojeili, C. 2015. Reflections on Wallerstein: The modern world-system, four decades on. *Critical Sociology* 41(4–5): 679–700. https://doi.org/10.1177/0896920513497377.

Erman, T. and Eken, A. 2004. The 'other of the other' and 'unregulated territories' in the urban periphery: Gecekondu violence in the 2000s with a focus on the Esenler case, Istanbul. *Cities* 21(1): 57–68. https://doi.org/10.1016/j.cities.2003.10.008.

Escobar, A. 2018. *Designs for the Pluriverse: Radical interdependence, autonomy, and the making of worlds*. Durham, NC: Duke University Press.

Eubanks, V. 2018. *Automating Inequality: How high-tech tools profile, police, and punish the poor*. New York: St Martin's Press.

Gabrys, J. 2014. Programming environments: Environmentality and citizen sensing in the smart city. *Environment and Planning D: Society and space* 32(1): 30–48. https://doi.org/10.1068/d16812.

Graham, M. 2019. Changing connectivity and digital economies at global margins. In *Digital Economies at Global Margins*, edited by M. Graham, 1–18. Cambridge, MA: MIT Press.

Graham, S. and Marvin, S. 2001. *Splintering Urbanism: Networked infrastructures, technological mobilities and the urban condition*. Abingdon: Routledge.

Guma, P. K. 2022. On tackling infrastructure: The need to learn from marginal cities and populations in the Global South. *Journal of the British Academy* 10: 29–37. https://doi.org/10.5871/jba/010.029.

Gururani, S. 2024. Paradoxes of the periphery. *Asian Geographer* 41(2): 143–50. https://doi.org/10.1080/10225706.2024.2340966.

Gururani, S., Kennedy, L. and Sood, A. 2021. *Special Issue: Engaging the urban from the periphery. South Asia Multidisciplinary Academic Journal* 26. https://doi.org/10.4000/samaj.7131.

Hauswedell, T., Körner, A. and Tiedau, U. (eds). 2019. *Re-Mapping Centre and Periphery: Asymmetrical encounters in European and global contexts*. London: UCL Press.

Heeks, R. and Shekhar, S. 2019. Datafication, development and marginalised urban communities: An applied data justice framework. *Information Communication and Society* 22(7): 992–1011. https://doi.org/10.1080/1369118X.2019.1599039.

Hernandez, D. O. and Titheridge, H. 2016. Mobilities of the periphery: Informality, access and social exclusion in the urban fringe in Colombia. *Journal of Transport Geography* 55: 152–64. https://doi.org/10.1016/j.jtrangeo.2015.12.004.

Hoefsloot, F. I. and Gateri, C. 2024. Contestation, negotiation, and experimentation: The liminality of land administration platforms in Kenya. *Environment and Planning D: Society and space* 42(5–6): 645–63. https://doi.org/10.1177/02637758241254943.

Hoefsloot, F. I., Martinez, J. and Pfeffer, K. 2022. An emerging knowledge system for future water governance: Sowing water for Lima. *Territory, Politics, Governance* 12(6): 825–45. https://doi.org/10.1080/21622671.2021.2023365.

Hoefsloot, F. I., Richter, C., Martínez, J. and Pfeffer, K. 2022. The datafication of water infrastructure and its implications for (il)legible water consumers. *Urban Geography* 44(4): 729–51. https://doi.org/10.1080/02723638.2021.2019499.

Hull, M. 2012. *Government of Paper: The materiality of bureaucracy in urban Pakistan*. Berkeley, CA: University of California Press.

Keil, R. 2017. *Suburban Planet: Making the world urban from the outside in*. Cambridge: Polity Press.

Kitchin, R. 2020. Civil liberties *or* public health, or civil liberties *and* public health? Using surveillance technologies to tackle the spread of COVID-19. *Space and Polity* 24(3): 362–81. https://doi.org/10.1080/13562576.2020.1770587.

Kitchin, R. and Dodge, M. 2011. *Code/Space: Software and everyday life*. Cambridge, MA: MIT Press.

Kohli, A. 2004. *State-Directed Development: Political power and industrialization in the global periphery*. Cambridge: Cambridge University Press.

Kwan, M-P. 2002. Time, information technologies, and the geographies of everyday life. *Urban Geography* 23(5): 471–82. https://doi.org/10.2747/0272-3638.23.5.471.

Masiero, S. and Das, S. 2019. Datafying anti-poverty programmes: Implications for data justice. *Information Communication and Society* 22(7): 916–33. https://doi.org/10.1080/1369118X.2019.1575448.

Mattern, S. 2021. *A City is Not a Computer: Other urban intelligences*. Princeton, NJ: Princeton University Press.

McCall, M. K., Napoletano, B. M., Boni Noguez, A. and Rico-Rodríguez, T. 2021. Territory in Latin America: An evasive and deeply embedded construct. In *Territorialising Space in Latin America*, edited by M. K. McCall, A. Boni Noguez, B. Napoletano and T. Rico-Rodríguez, 9–33. Dordrecht: Springer.

Milan, S. and Beraldo, D. 2024. Data in movement: The social movement society in the age of datafication. *Social Movement Studies* 23(3): 265–84. https://doi.org/10.1080/14742837.2024.2331550.

Petrusewicz, M. 2019. Rethinking centre and periphery in historical analysis: Land-based modernization as an alternative model from the peripheries. In *Re-Mapping Centre and Periphery: Asymmetrical encounters in European and global contexts*, edited by T. Hauswedell, A. Körner and U. Tiedau, 17–26. London: UCL Press.

Phelps, N. A., Maginn, P. J. and Keil, R. 2023. Centring the periphery in urban studies: Notes towards a research agenda on peripheral centralities. *Urban Studies* 60(6): 1,158–76. https://doi.org/10.1177/00420980221135418.

Pieterse, M. 2018. Where is the periphery even? Capturing urban marginality in South African human rights law. *Urban Studies* 56(6): 1,182–97. https://doi.org/10.1177/0042098018755067.

Schmid, C. 2018. Journeys through planetary urbanization: Decentering perspectives on the urban. *Environment and Planning D: Society and space* 36(3): 591–610. https://doi.org/10.1177/0263775818765476.

Scott, J. C. 1999. *Seeing Like a State: How certain schemes to improve the human condition have failed*. New Haven, CT: Yale University Press.

Shaw, J. and Graham, M. 2017. An informational right to the city? Code, content, control, and the urbanization of information. *Antipode* 49(4): 907–27. https://doi.org/10.1111/anti.12312.

Simone, A. 2010. *City Life from Jakarta to Dakar: Movements at the crossroads*. Abingdon: Routledge.

Simone, A. 2022. *The Surrounds: Urban life within and beyond capture*. Durham, NC: Duke University Press.

Simone, A. and Castán Broto, V. 2022. Radical unknowability: An essay on solidarities and multiform urban life. *City* 26(5–6): 771–90. https://doi.org/10.1080/13604813.2022.2124693.

Simone, A., Somda, D., Torino, G., Irawati, M., Niranjana, R., Bathla, N., Castriota, R., Megliò, S. and Chandra, T. 2023. Inhabiting the extensions. *Dialogues in Human Geography*. https://doi.org/10.1177/20438206231168896.

Sood, A. 2021. The speculative frontier: Real estate, governance and occupancy on the metropolitan periphery. *South Asia Multidisciplinary Academic Journal* 26: 71–99. https://doi.org/10.4000/samaj.7204.

Sriraman, T. 2018. *In Pursuit of Proof: A history of identification documents in India*. Oxford: Oxford University Press.

Star, S. L. and Ruhleder, K. 1996. Steps toward an ecology of infrastructure: Design and access for large information spaces. *Information Systems Research* 7(1): 111–34. https://doi.org/10.1287/isre.7.1.111.

Willett, J. and Lang, T. 2018. Peripheralisation: A politics of place, affect, perception and representation. *Sociologia Ruralis* 58(2): 258–75. https://doi.org/10.1111/soru.12161.

Wyatt, S. 2021. Metaphors in critical internet and digital media studies. *New Media & Society* 23(2): 406–16. https://doi.org/10.1177/1461444820929324.

Zook, M., Dodge, M., Aoyama, Y. and Townsend, A. 2004. New digital geographies: Information, communication and place. In *Geography and Technology*, edited by S. D. Brunn, S. L. Cutter and J. W. Harrington, 155–78. Dordrecht: Kluwer.

Theme I

Producing the informational periphery

2

Peripheralised labour: the digitisation of Nairobi's land information system

Ayona Datta and Dennis Mbugua Muthama

'I have not paid them,' the county official responsible for developing a digital land management database responded when we asked about the payment of casual workers involved in data extraction and digitisation activities (pers. comm. 2 July 2024). This remark came a full year after the completion of the work. The county in the peripheries of Nairobi had actively sought 'casuals', as they were called, to get its vast repositories of paper digitised as quickly and cheaply as possible. The delay in paying for services rendered is indicative of the precarious conditions under which casual workers operate within digital land administration systems. Their willingness to wait over a year for payment, without confronting the county government or the official in charge, shows their low bargaining power in this relationship. The fact that their payments have been delayed for over a year points to their invisibility – their location within the informational peripheries of state digitisation.

Kenya is in the midst of a push to digitise, with the Kenya Kwanza administration promising to digitise more than five thousand public government services (President Office 2022; *The Star* 2023). This drive is framed as a means to improve the efficiency of public service delivery, establish Kenya as a regional digital hub, and create a digital economy that empowers the country's youth to create employment opportunities (President Office 2022; President Office 2023a). By 30 June 2023, five thousand government services could be accessed online (President Office 2023b). In line with this national digitisation initiative, the digitisation of land administration services is being pursued.

In this chapter, we argue that Kenya's digitisation begins with the labour-intensive task of converting paper records in the land records department of each county into digital formats. Unless paper records are digitised, the information they contain cannot be transferred to county governments' digital platforms. Our aim is to understand how the digitisation of paper is operationalised through the embodied labour of casualised temporary workers inside the state. The transition from paper documents to digital platforms presents challenges on several fronts: curating vast repositories of paper records, documents, memos and titles across several thousand ledgers and case files; systematising a process of scanning and digitisation to capture the information held on paper in digital form; and training personnel to do this efficiently, accurately and transparently on a large scale. Significantly, while state officials are involved in this process (Datta and Muthama 2024), this flow of information from paper to digital primarily requires the integration of the 'peripheralised labour' of interns and casual workers into the workflow of state digitisation. Similar to the understanding of informational peripheries proposed in Chapter 1, the peripheral labour of the interns and casual workers discussed in this chapter is characterised by (1) invisibility due to purposeful ignorance on the part of officials, and (2) labour extraction in the service of urbanisation – in this case a digital land administration system.

Digitising paper records involves an elaborate human infrastructure that includes coding, categorising, scanning, verifying and uploading. How information flows between paper and digital records are conceived, implemented and administered by peripheral labour reflects how effectively land records are captured on a digital platform. In this chapter, we analyse how 'peripheralised labour' is embedded in the digitisation of paper records within the Nairobi Metropolitan Region, a process vital to upholding and strengthening the transition to digital governance. We argue that the 'peripheralised labour' of casual workers in the state is central to the politics of digitisation and land administration in the future. Even as the digitalising state engages in a 'distinct shift towards a technocratisation of urban governance' (Datta 2023), it requires a broad range of human labour at various scales and spaces within the state who are engaged in creating, regulating, monitoring and implementing its digitalisation initiatives.

Johnstone et al. note:

> Peripheral workers, also referred to as contingent workers, are typically those engaged in temporary part-time, fixed-term contracts, zero-hour contracts or casual employment. They may also include interns and temporary workers dispatched from employment agencies. (Johnstone et al. 2023, 306)

Here, we understand 'peripheralised labour' as the largely unrecognised and invisible embodied work of those often lowest in the hierarchy of state bureaucracy. Although much has been written in anthropology about state officials and their role in maintaining everyday paper bureaucracies (Das and Poole 2004; Hull 2012b; Hoag 2011; Gupta 2012; Mathur 2018), the role of peripheral workers in this process has largely been bypassed in the literature. This is a particularly significant omission in the digital age, as state capacity and resources for delivering digitalisation are limited, and this has increased the need for intervention from private-sector and non-state actors in delivering these goals. In a context where existing state officials do not have the necessary skills or are already overstretched with their workloads, interns and casual workers have tended to take on the banal work of digitising paper. They are peripheral in two senses: they remain outside state employment structures and pay scales, and the work they do is not considered central to state bureaucracy.

In Kenya, peripheralised labour is also defining the informational periphery of metropolitan regions such as Nairobi. Peripheralised workers constitute the shifting temporal shadow-bodies of the state, highlighting a significant informational gap in our understanding of the digitalising state. Peripheralised labour operates in a paradox – its workers remain in the informational peripheries of state bureaucracy, and yet they are central to the workings of this bureaucracy as it transitions to digital platforms. Peripheralised labour is temporal, casualised and largely precarious, and yet it provides essential training and pathways to more regular work for educated Kenyan youth. Peripheralised labour is understood as inconsequential to the bigger picture of a digitalising state, and yet state internship programmes attempt precisely to harness this labour in pivoting towards a digital future. Peripheralised labour is tied to the informational peripheries of the state and society as a geographic and virtual paradox of invisible labouring bodies at the centre of the digitalising state.

In this chapter we will focus on two types of peripheralised labour embedded within the digitalising state: (1) interns, who are recruited from a pool of university students seeking work experience; and (2) casuals, who are young Kenyans with tertiary education and able to undertake the unskilled labour of digitisation. Taken together, these two types of labouring bodies perform the bulk of paper digitisation, but there is also a clear hierarchy between them. While interns might in the future have the opportunity to pursue professional careers within state institutions, casuals do not necessarily have the same prospects. Both of these forms of labour enable informational flows from paper documents to the digital platform, and yet both are located within the informational peripheries of the state.

The 'peripheralised labour' of state digitisation

In recent years, there has been a growing interest in urban digital platforms designed to automate specific financial and property relations (Cowan 2021; Sadowski 2020; Mann and Iazzolino 2019; Kitchin et al. 2016). Scholars have observed that these platforms have actually produced higher levels of obscurity and illegibility on the ground than previously existed (Zook and Spangler 2022). While there has been much interest in FinTech and PropTech platforms, there is simultaneously an emerging interest in the processes of state digitalisation (Baud et al. 2014; Richter 2011) that are foundational to creating automated digital platforms. Indeed, land – as one of the key assets of the state – is now increasingly digitised, with the aim of automating the planning and governance of metropolitan regions. Yet, while producing detailed accounts of the power dynamics between state and private actors that digitalisation amplifies, and the disempowerment of communities that the digitisation of land often produces, the current scholarship is somewhat silent on the human labour required to transition to a fully automated aspirational future. Who carries out the digitisation work that is central to the digitalising state? How are records transferred from paper to digital platforms? Who contributes to the digitisation of land records, and what are the human impacts of these processes?

The answer to these questions can be found not just by scrutinising platforms, but also by moving behind the platforms to conduct a deeper examination of the human processes and practices that transfer paper records onto digital platforms. The processing of paper records, which includes methods of collecting, categorisation, storing and retrieving information related to specific land parcels, is maintained by a substantial state machinery of human labour. This labour includes not just state officials involved in land administration processes but also, crucially, casualised labour. These temporary workers are recruited to deal with specific parcels of work for a specified amount of time, often performing low-end tasks. This work is peripheralised in the wider narratives of a digitalising state, even as it begins to resemble a gig economy within state spaces.

In anthropologies of state bureaucracy, human labour is often depicted as that which is enacted by state officials in secure employment. For example, in his research on paper bureaucracies in Pakistan, Matthew Hull found that how paper is sorted, curated, stored and digitised reflects the political strategies and power dynamics of particular state officials and organisations at various scales (Hull 2012a). Similarly, in a detailed ethnography of health research in Malawi, Crystal Biruk found that quantitative data is entangled with the lives and experiences of the researchers contributing their labour in its collection, processing and storage (Biruk 2018). As Biruk noted, the staff who collect census data and create its categories are key actors in the technologies of governance, surveillance and control. In the same way, the processes of coding, categorising, sorting and scanning of paper for the automation of land administration in Kenya can also be seen as lying 'at the core of global governance regimes today' (Biruk 2018, 21). Yet, we know very little about the frontline workers involved in delivering Kenya's national aspiration of a digital economy.

In recent years, as work becomes increasingly digitised, scholars have drawn attention to the rise of the gig economy, which derives its name from the fact that each piece of work is treated as an individual performance or 'gig'. Often described as a feature of the 'fourth industrial revolution' (de Ruyter et al. 2018), the gig economy operates in a wide range of platforms across various sectors that hire independent contractors, consultants and workers, such as information technology, content creation, social media marketing, communications, food services, and creative fields such as art and design. It is a service-driven, technology-driven economic system in which workers are brought together on digital platforms, offered flexible hours but assigned specific, well-defined tasks. While this system claims to support flexibility of time and labour, it is more beneficial to capital than the worker, as the latter is not offered corresponding welfare or security measures.

While there is a growing body of literature on the gig economy, mainly as a result of the developing scholarship on labour geographies, concerns regarding gig work have largely emerged in the context of platformisation of labour involving Uber, Airbnb, Lyft and so on. Referred to as 'on-demand economy' (Todolí-Signes 2017), gig work creates an ambiguity in the definition of 'worker' as employee or self-employed. This, combined with the argument that some gig workers might themselves prefer the autonomy and flexibility of their work (Barratt et al. 2020), leads to their exclusion from the social protections offered by labour laws. Scholars in this field argue that those engaged in gig work should be recognised as 'workers' and, therefore, should be recognised within the purview of labour laws (Perritt 2019; Stewart and McCrystal 2019). Indeed, Stewart and Stanford (2017) present five main approaches to regulating gig work, from redefining the category of 'worker' to that of 'employer', and most importantly urging regulators to be 'creative and ambitious' when it comes to protecting the labour rights of gig workers.

For example, Rosenblat (2018) narrates how drivers in major US and Canadian cities navigate technology-driven work environments, dealing not only with the challenges of working conditions but also issues such as sexual harassment, racial discrimination and violations of labour rights. This has become even more significant in the context of COVID-19 lockdowns across the world, when recent scholarship has argued that gig workers should be seen as 'essential' or 'key workers' (Acevedo 2020).

Despite the growing richness and diversity of research on gig work, current scholarship on the gig economy – even in the context of the Global South – largely associates gig work with the coming of a digital revolution. In the Global South, a new digital economy has continued to extract traditionally undervalued labour into new spaces of 'informality, precarity, and exploitation' (Hunt and Samman 2020). For example, Hunt and Samman (2020) note that in the context of South Africa, domestic workers have been ensnared in new on-demand platform models for delivering their services. Woodcock and Graham's (2020) work on geographically tethered work and cloud work unravels the ethnographic details of platforms such as Ola and Uber in the UK, Ghana, South Africa and India. They argue that digital infrastructures – such as internet access and GPS – play an important role in enabling corporates to maintain dual marketplaces through consumer preference for on-demand services, leading to the prioritisation of corporate profit over workers' rights. Berg et al. (2018) differentiate between gig workers on web-based platforms (freelancers, micro-tasking and content-based creative-crowd-work) and those on location-based platforms (transportation, delivery, household services). Ravenelle (2019) explores the daily challenges of contractors working on platforms like TaskRabbit, Uber, Kitchensurfing and Airbnb, highlighting the struggles they face sustaining the kinds of jobs that are available in the wake of the 2008 financial crisis. Similarly, Cant (2019) offers an auto-ethnographic narrative as a part-time Deliveroo worker struggling to meet the daily cost of reproduction of labour.

In current literature on gig work, while the precarious nature of the work is much debated, the temporal vulnerabilities of gig workers are largely unacknowledged. This enables employers to misclassify workers, engage in 'regime shopping' and exploit the most vulnerable (Zwick 2018). It is also noted that time remains one of the key sites of negotiation and trade-offs among women service workers, who may be digitally inclined to make full use of the flexibilities of the platform but are unable to keep up with the rhythm and pace of the atomised work due to their domestic and socially reproductive roles (Datta 2020). As Surie and Koduganti (2016) find in the context of Uber and Ola drivers in India, a relational sense of time is central to their leaving full-time work and entering the time-space of piecemeal work. As they suggest, seeing gig workers through the lens of temporality enables us to see them as akin to informal workers – on which so much scholarship already exists. While the question of whether gig work is piecemeal, on-demand and therefore exploitative is widely debated, the fact that gig workers are also present within state spaces doing the piecemeal work of digitisation is rarely acknowledged.

In Kenya, the literature on the gig economy shows similar biases. The rich scholarship on gig workers largely focuses on Uber, Safaricom and other providers outside the state. The state is often seen as a sealed institution of secure employment. In his ethnography of Uber workers in Nairobi, Iazzolino states that workers are often able to game the platform by collectively shaping demand and supply, thus highlighting their collective agency in economic spaces (Iazzolino 2023).

Overall, then, in Kenya and the Global South, the gig economy remains focused on the private sector of service work and has largely overlooked the prevalence of casualised labour within the state. State employment is often seen as secure and well paid, offering several employment benefits. One of the reasons why casualised labour within the state has remained peripheral to the gig economy is because it is well hidden within the banal spaces of the everyday state. In Kenya, because of its internship programme and various other state initiatives to include unemployed youth in mainstream work, precarious labour within internships is not usually considered gig labour.

What is 'peripheral labour'? Cowan et al. argue that this concept 'recasts labor in the "peripheries", not as an externalised quantity redundant to emerging economic formations, but rather as integral if often hidden features of capitalist value extraction' (Cowan et al. 2023, 7). Drawing on a Marxist approach of accumulation by dispossession, they argue that peripheral labour is surplus labour that is extracted through its simultaneous devaluation. Peripheral labour exists even within the core of the formalised, 'so-called Standard Employment Relation' (Cowan et al. 2023, 8). As Mishra notes in the case of marginal workers in brick kilns, this sort of labour is both geographic and temporal, pushed to the peripheral spaces of metropolitan regions while negotiating the temporalities of capital flows in the centre (Mishra 2021). There is a need, then, to focus on this peripheral labour if we are to understand the land administration machine as an apparatus of state power and sovereignty. If platforms construct an automated future of the state, it is human labour that makes this aspirational future possible.

We argue that peripheral labour should be rethought as an ongoing relational process, where precarious labour conditions continue to persist in the margins of an information age. In this chapter we use the notion of peripheralised labour as a process of creating a changing relationship with the digitalising state through two components: time and data. Time refers to both the speed of labour and the speed of data flows. Here we argue that time is only measured when it benefits the state, not – as we noted at the start of this chapter – when it delays the remuneration of casual workers. Similarly, data too is measured in terms of the speed of flows from paper to platform, parcelled into micro-work to extract maximum value from peripheralised labour. Peripheralised labour, then, is an embodied socio-technical relationship between young precarious workers and the apparatus of the state.

Researching peripheral labour in the state

This chapter was written as part of the ongoing 'Regional Futures' project, which is investigating the dynamic of digitalisation-as-urbanisation in the peri-urban municipalities of three rapidly growing metropolitan regions: Mumbai (India), Nairobi (Kenya) and Guadalajara (Mexico). In the Kenyan context, the chapter reports on research being undertaken in two peripheral counties located in the Nairobi Metropolitan Region.

Kenya has two levels of governance: national and county. Land administration and management are undertaken at both levels. The land digitisation practices described in this chapter are part of the ongoing implementation of the National Land Information Management System, commonly known as Ardhisasa. This system is being implemented by the national government, with support from key stakeholders including county governments, land professionals and civil society actors. It is the experiences of the interviewees with the development and use – or lack thereof – of this system that this chapter presents as land digitisation practices within the Nairobi Metropolitan Region.

The arguments in this chapter are developed from 22 semi-structured interviews conducted between November 2022 and February 2023 with national and county government officials, civil society actors, land professionals and land brokers. The participants, drawn from three of the five Nairobi metropolitan counties, were either directly involved in some aspect of the land digitisation programme or had first-hand knowledge and experience of documenting its practices or specific aspects of it. The interviews have been augmented using additional data from further interviews conducted between March 2023 and July 2024 with county government officials.

Prior to undertaking interviews with officials at the two levels of governance, research permission and authorisation were obtained from the national body – the National Commission for Science, Technology and Innovation. After receiving the research permit, formal approval to interview and access information was sought from the leadership of various agencies and departments at both national and county levels. Throughout this process, the research team benefited from good working relationships with national and county government officials and land professionals, fostered through previous professional and research undertakings.

The fieldwork comprised interviews with informants who had knowledge of the land digitisation programme. We were introduced to these participants through professional and academic networks, which acted as their gatekeepers. Given that land is an emotive issue in Kenya and that security institutions were involved in the digitisation programme, potential participants were understandably hesitant and cautious when it came to discussing the project. However, following trust-building conversations and assurances of confidentiality and privacy, they agreed to engage in the research. Following these initial interviews, we adopted a snowballing method through which we recruited other participants.

Our diverse positionalities as Kenyan and UK-based researchers fostered different relationships with participants. The Kenyan researcher, who had previous experience working on land issues in Kenya, was an 'insider' as they understood many of the land issues that were being discussed. However, they were also, to some extent, an 'outsider' due to their ethnic/tribal background. This nuanced duality of 'insider' positioning through research experience and 'outsider' status due to their ethnic background informed field observations and the field notes. For the UK researcher – a complete outsider in the Nairobi context – it was notable that participants often took time to elaborate on the local context, which in the case of the Kenyan researcher was taken as a given. We kept detailed accounts of these encounters in our fieldnotes, and they have enriched our analysis in this chapter.

The Kenyan youth unemployment crisis

In recent years, the Kenyan government has launched several initiatives aimed at increasing youth employment in response to high unemployment rates and as part of its broader strategy to achieve the goals set out in the Kenya Vision 2030 policy. One of these is the Skill Development and Comprehensive Training programme, which seeks to 'enhance the professional development of young people and prepare them for the labour market, to expose them to technical skills that are pivotal both in the formal and informal sector' (Mwaniki and Omariba 2024). Its notable initiatives include the National Youth Service (NYS) programme, the Kenya Youth Employment Opportunities Project, and the national Public Service Commission (PSC) internship programme.

The PSC internship programme stands out as a significant contributor to state digitisation. Launched in 2019 through the Kenyan Youth Policy (Republic of Kenya 2019), the programme was designed to address both the UN's Sustainable Development Goals 2030 and the concerns raised by the World Bank regarding the high levels of youth unemployment in Kenya, highlighted in 2016. It was also linked to the swathe of initiatives that were launched by the Kenyan government as part of its Vision 2030 plan.

The official definition of 'intern' in Kenya is 'a student or graduate with relevant qualifications who has entered into a contract with an organisation for a period of between three and twelve months with the intent of gaining supervised practical experience for registration with respective professional bodies and/or to increase chances of employability' (Republic of Kenya 2019, v). Prospective interns may apply directly to the Ministry or through the PSC. Where an applicant applies through the PSC, if they are successful they are posted to a national government agency or state department, such as the State Department of Lands and Physical Planning. The posting period is one year, and interns are paid a stipend of Kshs.25,000. Currently, this is the main route through by interns are placed in national government agencies. While the direct application route is also an option, according to one of our interviewees in the state department (pers. comm. 2 July 2024) only relatives of powerful individuals are successful through this route.

While the PSC operates at a national level, there are also separate county-level internship programmes operated by the county executive in line with the Kenyan Youth Policy 2019. At the county level, a person who is interested in an internship with one of the county departments, such as Lands and Physical Planning, applies directly to the chief officer in charge of the department. The chief officer will then write to the County Public Service Board requesting official deployment. In some cases, applications may arise on the recommendation of a person known to the official concerned, so it is not surprising to find interns deployed on the basis of who they know. Once the application is successful and the deployment is confirmed, an intern will join the department for three months. However, it is possible to extend this period beyond the three months – particularly for graduates, who may have the flexibility to remain longer.

Gathongo notes that interns present an inherent contradiction, since 'the true legal status of a person [intern] who is an employee is disguised in a way that hides his or her true legal status or gives it an appearance of a different legal nature' (Gathongo 2021, 1). This has been evident in recent events and public debates around the sustainability of the PSC internship programme, which aims to train around eight thousand young, skilled Kenyans within state governance spaces in a variety of areas including ICT, project management and research (Republic of Kenya 2019). While this target is ambitious, there are still far more young people applying for internships than there are vacancies. The Kenya Institute for Public Research and Analysis reports that the programme tends to be biased towards non-degree holders except in the area of ICT, where more advanced skills are needed (Mwaniki and Omariba 2024). There are also limited post-internship paid employment opportunities that interns can tap into (Gathongo 2021), although there have been recent reports of PSC interns occupying government offices to protest about their joblessness despite having successfully completed internship placements (*The Star* 2024).

Invisible labour/technical bodies

Land in Kenya has long been a highly politically charged context, as it has been connected to land grabbing, corruption, embezzlement and other extra-legal practices (Manji 2014). Land digitisation was seen as a national security concern, and therefore its implementation was managed at a national level. Due to the high-security nature of this task (stemming from suspicions that officials were manipulating paper documents), the planning and management of this process was undertaken by the Kenyan military and other national security institutions. As one official explained:

> They give you the physical files, then they give you the computer from which to work, you log in using the details provided through the facilitation of the ministry, the ministry would before that create a list of people authorised for a specific task, and share this with JNAM [Joint National and Resource Mapping] who would then generate your access code. The access code would be for that process only. (NA230217I002)

In two Nairobi counties where digitisation has taken place, the process was immensely space, time and labour intensive. Land records occupy significant physical space, requiring large facilities for storage and

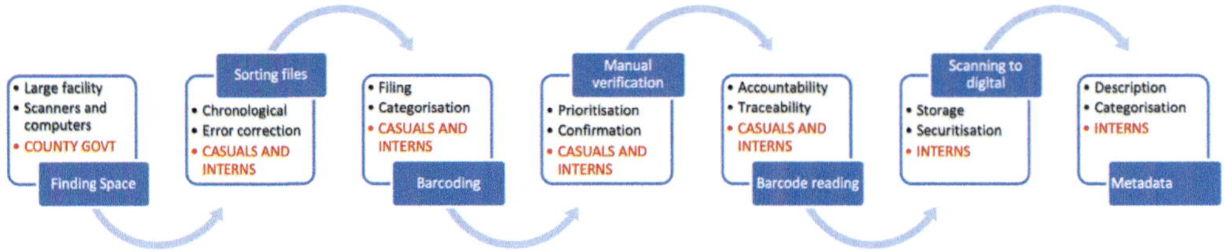

Figure 2.1 Flowchart showing the digitisation process in Nairobi counties.
Source: created by the authors.

cataloguing. Similarly, the digitisation process demands physical space – large office facilities with scanners, computers, servers and desk-space. The process is meticulous and involves sorting paper files, barcoding each piece of paper, manual verification, barcode reading, and scanning into digital format.

Digitisation was carried out in two phases: first, creating a system for tracking and tracing the paper documents; and second, implementing an Electronic Data Management System that was used to scan all the documents onto a computer. The role of state officials primarily involved the first phase. Once they had sorted, curated, recorded and verified the paper records, they had to hand the papers to the military. The military then recruited interns and casual workers to do the actual scanning and digital conversion. These workers were of a younger generation, with basic IT skills, and were not familiar with the land administration system. The process of scanning took months and state officials were kept at arm's length; those we spoke to described the process as being shrouded in secrecy.

Figure 2.1 shows the main steps involved in digitisation. Interns and casuals (as they were referred to) are largely hidden in this process of digitisation, although they are necessary at each stage of the process. The flowchart shows the stages at which casuals and interns are involved: the former take on those aspects of the digitisation process requiring minimal skill, while the latter are involved in all stages. Interns are particularly involved in the final two stages of scanning and metadata, as these require a higher degree of skill in securitisation and cataloguing of the data.

Flexible work/technical knowledge

The internships offered in county government departments primarily target recent graduates with technical skills and a flexible approach to work. In one county, the deputy director of ICT explained to us that interns are posted to specific departments after applying through the human resources department. Once placed, they are assigned to their respective directorates – for example, ICT interns are posted to the ICT directorate and report to the county's director of ICT. Through this structure, interns are involved in various aspects of the county's workflows, experiencing a diverse range of tasks and gaining the necessary experience for future employment. In practice, though, as the deputy director noted, the internship role is best understood as assisting with whatever is needed. This may range from actual technical work to more mundane responsibilities, such as photocopying documents or even going to the supermarket to fetch supplies. Furthermore, according to a physical planner in another county department, interns in the county are not usually paid. However, they may receive a stipend if they participate in fieldwork alongside officials. On a day-to-day basis, then, interns are expected to be flexible and provide just-in-time labour for their supervisors without the prospect of fair remuneration. Upon completing their internship, they are given an official acknowledgement letter.

Interns applying for positions in technical departments, such as the State Department of Lands and Physical Planning, are required to have a degree in the relevant field. As a result, they are expected to assist officials with some of the technical aspects of the department's work. Further, training is not necessarily provided during the internship, since they are expected to be able to do jobs in fields they already have a degree in, and to learn on the job. The quality of an intern's experience in this context is primarily dependent on the willingness of officials to train them and the availability of learning opportunities. In the Lands department, workloads were particularly heavy and officials did not have time to train interns, so they recruited interns who were land specialists who could then do the work with minimal supervision. One county official remarked that the county is not a good place for interns to gain experience, as its departments are not as dynamic as private companies, where there may be a number of projects being undertaken simultaneously.

The county official we interviewed was, however, of the view that interns gain more technical experience and have a brighter future than casuals. The difference between casuals and interns is that the latter are involved in or undertake technical land administration operations, while for the former, their work includes both technical and non-technical tasks. Casuals are employed on short-term, temporary contracts, usually lasting three months with no guarantee of renewal. They are usually paid Kshs.1,000 per day and are recruited from local towns. The main qualifications required for casuals are at least a tertiary education qualification and basic computer skills. In one county department, the official in charge of data extraction from maps to develop a digital list of land parcel owners in the county's urban centres noted that he used 'casuals' to do this work. He employed 20 casuals who reported to him and signed a daily attendance form. The casuals were working in a borrowed space in the boardroom of the Directorate of Revenue.

The official informed us that in the digitisation project they seek more skilled casuals or interns since the work is short-term, meaning the county can get the work done more quickly and at a cheaper rate.

> [Y]ou get 20 to 15 casuals depending on the amount of work that is needed. And also, the people who man those current records must be in the mix. So it is usually around 25 people ... The casuals are [there] because you can't employ somebody permanent or give somebody a contract to do that. And then when it ends, what does he do? (NA230731I034)

The nature of this flexible, just-in-time labour and its temporariness produces a culture where casuals can be asked to undertake intense workloads for extended periods. As explained by an official in one state department which was more interested in hiring casuals than interns:

> That's why we ask for casuals. And this is the wrong way. But these guys will do anything. They want money. So, he's there early, he does a lot of work. Leaves late. You get? Yeah. (NA230731I034)

In this description of casual labour, the flexible nature of on-demand labour is arguably analogous to gig work, but also distinctly different from it. Like gig workers, casuals and interns do exist somewhat within a 'just-in-time economy' (Zwick 2018) of state digitisation. However, their position is not quite so precarious, as they are technically better educated and more skilled than those working in service-sector platforms such as Uber. While both groups face challenges, with low pay and little control over their working conditions, casuals and interns are peripheral labour relative to the labour of state officials in the digitalising state.

Parcelling digitisation into micro-work

An important aspect of using interns and casuals in the workflows of digitisation was to reduce costs and ensure smooth data capture. But this also meant parcelling the data into work packages that could be delivered independently by temporary, flexible labour.

As one official explained, his department used interns to work on its land information management system. However, due to the temporary nature of the labour available from the interns, the department was unable to entrust them with overseeing the entire project. As a result, the data was parcelised and broken up into smaller work packages, which the interns could complete within their limited time in the department. This occurred in a context where the software was outdated, and expensive proprietary software such as ArcGIS was not feasible, so the interns had to use open-source software. The challenges such conditions create for interns, learning to negotiate new software and adapting to unfamiliar working conditions, go largely unacknowledged by county officials. More crucially, it is evident that parcelising data goes against the ideology of seamless, interoperable and ubiquitous data transfers envisioned for the platform.

> We were to now check this is the file and this is the record that has been captured. Through the process of data capture by interns and junior officers and the NYS characters who won't understand. But you need a land expert. (NA240613219S)
>
> These are the processes, the expert processes. These are the resource officers then the data capture, these are ... only one, especially NYS, interns and all that. And the eight experts, these are highly trained characters, expert in programming. (NA240613219S)

Due to the parcelisation of data capture from paper to the platform, county officials were required to undertake quality assurance and check for consistency across the work of the interns. This created a clear hierarchy: even though some of the interns were as well qualified as land experts, due to the fragmented nature of the work they did not have a broader overview. As a result, they were kept at the bottom of the hierarchy, working for the county land officers. This fragmented and parcelised model is similar to how the gig economy captures just-in-time micro-work to reduce risk, increase speed and create further precarity of labour. Thus, interns were embedded in the platform preparation mainly for data capture, as larger packages of work required managerial responsibilities that could only be undertaken by permanent employees.

One of our interviewees was a private software developer. He felt that this parcelisation of digitisation into micro-work, involving what was seen as a 'core' element of state technology, was disappointing. The developer argued that the state was not prioritising the quality of workers in the digitisation process as they were being used mainly for quick fixes. This, he noted, was problematic as interns were doing the work that ought to be done by fully employed professionals, and this perpetuated – in his view – half-measure solutions – or being 'pushed along'.

> Yes, fixing the network. So they don't employ creme de la creme. They just look for interns to just let it continue pushing the [work] ... So that needs to be pushed along. I don't know, ICT is being seen as a core. (NA231201I044)

If interns were low in the hierarchy of labour in the digitisation process, casuals were particularly peripheralised within peripheral labour. One of the county officials noted that they generally used casuals to do the work of interns – specifically, students from a local university on break. Using them as casuals meant that they could get the 'unskilled' labour of entry-level digitisation completed in 'no time' through a bulk hire of 100 students.

> Then [the Director] got students from [university], interns. Well, he just picked students, I think he was paying them something per day. They were over 100 students, just to scan the document and digitise on ArcGIS. Digitise the boundaries. Within no time, we had all the parcel boundaries for [neighbourhood] ... Yeah. So it's able to sustain itself. And you see, he was engaging interns. Interns, actually students, not interns. They were students ... during the long holiday. You engage them. It's just a small fee. They are cheap. You know they appreciate. (NA240514I062)

Casuals, then, fulfilled two crucial requirements of state digitisation – speed and affordability – without the need for formal employment. However, this made their work further parcelised, for example in relation to executing micro-tasks like 'digitising boundaries' on the geospatial platform. It required minimal commitment from the casuals as they were doing micro-work during a temporary break from their studies. For everyone it seemed like a 'win-win' situation, although, as we noted at the beginning of this chapter, timely and reliable payment was never a guarantee for the casual workers.

Planning a future with peripheral labour

Given the centrality of internships and casualised labour in state digitisation, future plans for digital resources (such as geographic information systems (GIS) or computer labs) were often based solely on the availability of interns and casuals. In the view of the county officials we interviewed, the work was made easier by using interns, as well as other cost-effective resources such as open-source software. For example, due to its ease of access and transferability, most of the interns' digitisation tasks were carried out on QGIS, an open-source geospatial software, rather than proprietary software such as ArcGIS. Much like open-source software, peripheralised labour was easier to access and faster in terms of delivery of work. This meant that much of the future of state digitisation was planned around open-source software and casualised labour, as the processes of hiring permanent staff and requisitioning new proprietary software from state funds were too slow and too expensive for the kinds of jobs that were required for digitisation.

> I don't even need to wait for a donor or for money to be provided so that I can be able to do it. I can use an open-source solution like QGIS, right? I can have interns here in the office, it is you who will digitise that data and I have a base. When you bring in your work, I will see to it that it is actually being put where it needs to sit. So, it's just, I don't know. These things are not complex. (NA230314I006)

In one of our interviews, the County Director of Survey noted that, following the construction of a new GIS lab, he will use interns to digitise most of the county records. Here the main obstacle was seen to be physical space rather than human labour, as this latter aspect of digitisation was readily provided by interns and casuals.

> Because now I am waiting for that space to be there. So as you can now concentrate on having interns to do the whole. (NA230418I012)

The future of a digitalising state does present opportunities for interns to transition to permanent employment, though whether they are offered such opportunities often depends on the knowledge and expertise they are able to demonstrate during their brief internship. County officials noted that the best interns are subsequently hired, as their experience is valuable for the future of state digitisation:

> We get interns to come and help us. We get interns. People who are knowledgeable in those fields. And actually, when we started this five years ago, most of those who had been co-opted have now been incorporated and been employed for continuity. (NA231103I041)

Here we see the stratified structure of precarious labour: some interns might eventually have a permanent future within the state, while casuals can never harbour such hopes. Peripheralised labour is not just precarious surplus labour located in the informational peripheries of the state; it is inherently uneven, differentiated and hierarchical, whereby some workers are able to find their way to the centre while others will always remain in the margins.

State digitisation as precarious labour

In this chapter, we have used the concept of peripheral labour – an ongoing relational process characterised by precarious labour conditions in the margins of the digital state – to argue that digitisation requires vast reserves of casualised peripheralised workers and produces new kinds of precarity, and ultimately devalues surplus labour. Digitisation claims to make the work of state officials more efficient; however, this comes at the expense of unrecognised, undervalued and invisible labour performed by those outside the state. The interns and casual workers who do the bulk of paper digitisation may be formally located outside state spaces, yet they run the machines of digitisation within the state so that it can plan for an automated future.

Thus, we have argued that the peripheralised labour of young, unemployed Kenyans constitutes the informational periphery of state digitisation projects, despite the fact that they play a central role in entrenching and supporting Kenya's future transition to an automated digital land administration platform. The prioritisation of speed over process in digitisation means that it is possible to see state digitisation as an extraction of the precarious time of its workers – in terms of both working beyond official hours and waiting for delayed payment. This dynamic enables the state as employer(s) to accumulate more work from peripheralised workers on demand, making the labour flexible and stretching it beyond standard working hours. It coerces young Kenyans, particularly those with tertiary education, to make difficult trade-offs between their various time-bound duties at the very least, and sacrifices of personal time and long-term economic development at the worst.

While informal labour has always been precarious and uncertain across the Global South and North, its institutionalisation within the state and its normalisation as youth employment extend this precarity in the future. On the one hand, employment within the state promises a more secure future for young Kenyans, offering valuable experience and transferable skills. On the other hand, on-demand work, with its associated wage uncertainties, increases uncertainty in planning for the future of state digitisation initiatives. The parcelised, atomised on-demand labour that is currently fuelling state aspirations for digitisation has in-built faultlines that extend the crisis of livelihood, health and welfare for young Kenyan workers.

Acknowledgements

An earlier version of this chapter was published as A. Datta and D. M. Muthama, Sorting paper: The archival labour of digitising land records in Kenya. (2024) *The Geographical Journal* 190(4). https://doi.org/10.1111/geoj.12581.

Funding for this research was provided by the European Research Council Advanced Grant 'Regional Futures: The territorial politics of digitalisation-as-urbanisation in the Global South' (ID: 101019318) awarded to Ayona Datta. We would like to thank the land administration officials based in various Nairobi counties for their cooperation and support and for patiently answering our questions. We would also like to thank our project team colleagues for their feedback on earlier versions of this chapter.

References

Acevedo, Deepa Das (ed.). 2020. *Beyond the Algorithm: Qualitative insights for gig work regulation*. Cambridge: Cambridge University Press.

Barratt, Tom, Caleb Goods and Alex Veen. 2020. 'I'm my own boss …': Active intermediation and 'entrepreneurial' worker agency in the Australian gig-economy. *Environment and Planning A: Economy and space* 52(8): 1643–61. https://doi.org/10.1177/0308518X20914346.

Baud, Isa, Dianne Scott, Karin Pfeffer, John Sydenstricker-Neto and Eric Denis. 2014. Digital and spatial knowledge management in urban governance: Emerging issues in India, Brazil, South Africa, and Peru. *Habitat International* 44: 501–9. https://doi.org/10.1016/j.habitatint.2014.09.009.

Berg, Janine, Marianne Furrer, Ellie Harmon, Uma Rani and M. Six Silberman. 2018. *Digital Labour Platforms and the Future of Work: Towards decent work in the online world*. Geneva: International Labour Organization.

Biruk, Crystal. 2018. *Cooking Data: Culture and politics in an African research world*. Durham, NC: Duke University Press.

Cant, Callum. 2019. *Riding for Deliveroo: Resistance in the new economy*. Cambridge: Polity Press.

Cowan, Thomas. 2021. Uncertain grounds: Cartographic negotiation and digitized property on the urban frontier. *International Journal of Urban and Regional Research* 45(3): 442–57. https://doi.org/10.1111/1468-2427.13016.

Cowan, Tom, Stephen Campbell and Don Kalb. 2023. Theorizing peripheral labor: Rethinking 'surplus populations'. *Focaal* 97: 7–21. https://doi.org/10.3167/fcl.2023.970102.

Das, Veena and Deborah Poole (eds). 2004. *Anthropology in the Margins of the State*. Santa Fe, NM: SAR Press.

Datta, Ayona. 2020. The 'smart safe city': Gendered time, speed, and violence in the margins of India's urban age. *Annals of the American Association of Geographers* 110(5): 1,318–34. https://doi.org/10.1080/24694452.2019.1687279.

Datta, Ayona. 2023. The digitalising state: Governing digitalisation-as-urbanisation in the global south. *Progress in Human Geography* 47(1): 141–59. https://doi.org/10.1177/03091325221141798.

Datta, Ayona and Dennis Mbugua Muthama. 2024. Sorting paper: Archival labour of digitising land records in Kenya. *The Geographical Journal* 190(4). https://doi.org/10.1111/geoj.12581.

de Ruyter, Alex, Martyn Brown and John Burgess. 2018. Gig work and the fourth industrial revolution: Conceptual and regulatory challenges. *Journal of International Affairs* 72(1): 37–50.

Gathongo, Johana Kambo. 2021. Internship: A bridge to employment or a trap to a disguised employment relationship? *Journal of Conflict Management and Sustainable Development* 7(1): 1–22. Accessed February 2025. https://journalofcmsd.net/wp-content/uploads/2021/08/Internship.pdf.

Gupta, Akhil. 2012. *Red Tape: Bureaucracy, structural violence, and poverty in India*. Durham, NC: Duke University Press.

Hoag, Colin. 2011. Assembling partial perspectives: Thoughts on the anthropology of bureaucracy. *PoLAR: Political and Legal Anthropology Review* 34(1): 81–94. https://doi.org/10.1111/j.1555-2934.2011.01140.x.

Hull, Matthew. 2012a. Documents and bureaucracy. *Annual Review of Anthropology* 41: 251–67. https://doi.org/10.1146/annurev.anthro.012809.104953.

Hull, Matthew. 2012b. *Government of Paper: The materiality of bureaucracy in urban Pakistan*. Berkeley, CA: University of California Press.

Hunt, Abigail and Emma Samman. 2020. Domestic work and the gig economy in South Africa: Old wine in new bottles? *Anti-Trafficking Review* 15: 102–21. https://doi.org/10.14197/atr.201220156.

Iazzolino, Gianluca. 2023. 'Going Karura': Colliding subjectivities and labour struggle in Nairobi's gig economy. *Environment and Planning A: Economy and space* 55(5): 1114–30. https://doi.org/10.1177/0308518X211031916.

Johnstone, Stewart, Jenny K. Rodriguez and Adrian Wilkinson (eds). 2023. *Encyclopedia of Human Resource Management*. 2nd edn. Cheltenham: Edward Elgar.

Kitchin, Rob, Sophia Maalsen and Gavin McArdle. 2016. The praxis and politics of building urban dashboards. *Geoforum* 77: 93–101. http://dx.doi.org/10.2139/ssrn.2608988.

Manji, Ambreena. 2014. The politics of land reform in Kenya 2012. *African Studies Review* 57(1): 115–30. https://doi.org/10.1017/asr.2014.8.

Mann, Laura and Gianluco Iazzolino. 2019. See, nudge, control and profit: Digital platforms as privatized epistemic infrastructures. Bangalore: IT for Change. Accessed February 2025. https://itforchange.net/platformpolitics/wp-content/uploads/2019/03/Digital-Platforms-as-Privatized-Epistemic-Infrastructures-_5thMarch.pdf.

Mathur, Nayanika. 2018. *Paper Tiger: Law, bureaucracy and the developmental state in Himalayan India*. Cambridge: Cambridge University Press.

Mishra, Pratik. 2021. The making of urban peripheries and peripheral labor: Brick kilns and circular migration in and beyond Greater Delhi. *South Asia Multidisciplinary Academic Journal* 26, 100–21. https://doi.org/10.4000/samaj.7276.

Mwaniki, Joseph and Vivian Omariba. 2024. Rethinking youth employment initiatives in Kenya: Lessons and policy alternatives. Kenya Institute for Public Policy Research and Analysis, 8 January. Accessed February 2025. https://kippra.or.ke/rethinking-youth-employment-initiatives-in-kenya-lessons-and-policy-alternatives.

Perritt Jr, Henry H. 2019. Don't burn the looms: Regulation of Uber and other gig labor markets. *SMU Science & Technology Law Review* 22(1): 51–149. Accessed February 2025. https://scholar.smu.edu/cgi/viewcontent.cgi?article=1289&context=scitech.

President Office. 2022. President Ruto: Government to digitise its operations. Republic of Kenya. Accessed February 2025. https://www.president.go.ke/president-ruto-government-to-digitise-its-operations.
President Office. 2023a. President Ruto: We must be efficient in service delivery. Republic of Kenya. Accessed February 2025. https://www.president.go.ke/president-ruto-we-must-be-efficient-in-service-delivery.
President Office. 2023b. President Ruto unveils online government services. Republic of Kenya. Accessed February 2025. https://www.president.go.ke/president-ruto-unveils-online-government-services.
Ravenelle, Alexandrea. 2019. *Hustle and Gig: Struggling and surviving in the sharing economy*. Berkeley, CA: University of California Press.
Republic of Kenya. 2019. *Kenya Youth Development Policy 2019*. Accessed February 2025. https://youth.go.ke/wp-content/uploads/2020/11/Kenya-Youth-Development-Policy-2019-Popular-version.pdf.
Richter, Christine. 2011. In-tensions to infrastructure: Developing digital property databases in urban Karnataka, India. *Environment and Urbanization Asia* 2(2): 205–22. https://doi.org/10.1177/097542531100200205.
Rosenblat, Alex. 2018. *Uberland: How algorithms are rewriting the rules of work*. Berkeley, CA: University of California Press.
Sadowski, Jathan. 2020. Cyberspace and cityscapes: On the emergence of platform urbanism. *Urban Geography* 41(3): 448–52. https://doi.org/10.1080/02723638.2020.1721055.
The Star. 2023. State has digitised 300 services ahead of June deadline – Ruto. *The Star*, 6 January. Accessed February 2025. https://www.the-star.co.ke/news/2023-01-06-state-has-digitised-300-services-ahead-of-june-deadline-ruto.
The Star. 2024. PSC interns threaten to occupy commission offices on Monday. *The Star*, 28 July. Accessed February 2025. https://www.the-star.co.ke/news/2024-07-28-psc-interns-threaten-to-occupy-commission-offices-on-monday.
Stewart, Andrew and Shae McCrystal. 2019. Labour regulation and the great divide: Does the gig economy require a new category of worker? *Australian Journal of Labour Law* 32. Accessed February 2025. https://ses.library.usyd.edu.au/bitstream/handle/2123/32829/Version%20of%20Record.pdf?sequence=1&isAllowed=y.
Stewart, Andrew and Jim Stanford. 2017. Regulating work in the gig economy: What are the options? *Economic and Labour Relations Review* 28(3): 420–37. https://doi.org/10.1177/1035304617722461.
Surie, Aditi and Jyothi Koduganti. 2016. 'The emerging nature of work in platform economy companies in Bengaluru, India: The case of Uber and Ola cab drivers'. *E-Journal of International and Comparative Labour Studies* 5(3). http://ejcls.adapt.it/index.php/ejcls_adapt/article/view/224.
Todolí-Signes, Adrián. 2017. The 'gig economy': Employee, self-employed or the need for a special employment regulation? *Transfer: European Review of Labour and Research* 23(2): 193–205. https://doi.org/10.1177/1024258917701381.
Woodcock, Jamie and Mark Graham. 2020. *The Gig Economy: A critical introduction*. Cambridge: Polity Press.
Zook, Matthew and Ian Spangler. 2022. A crisis of data? Transparency practices and infrastructures of value in data broker platforms. *Annals of the American Association of Geographers*, July: 1–19. https://doi.org/10.1080/24694452.2022.2071201.
Zwick, Austin. 2018. Welcome to the gig economy: Neoliberal industrial relations and the case of Uber. *GeoJournal* 83(4): 679–91. https://doi.org/10.1007/s10708-017-9793-8.

3

Waste as an economic and informational resource: the datafication of waste collection labour at India's informational periphery

Josie Wittmer

This chapter explores the production of the informational periphery by investigating labour experiences in India's contemporary solid waste management landscape. In the context of the current push for technological solutions to waste production in cities, I focus on the increasing datafication of human labour that is produced by the development and use of platforms, sensors and other information technologies for collecting and transporting garbage. Drawing on empirical examples from the city of Ahmedabad and analyses of 'smart' waste management practices emerging in other Indian cities, the chapter identifies compounding tensions between parallel processes of informational surveillance and exclusion experienced by waste workers in a sector currently dominated by public–private partnership (PPP) models and a push toward the mechanisation and digitisation of service delivery. I focus on data–power relations and the social consequences associated with these emerging forms of collecting waste and monitoring workers in order to situate a theorisation of the informational periphery within regimes of caste, capital and migration in India. In doing so, I argue that informational peripheries are continually produced through selective processes of datafication and surveillance, resulting in the entrenchment of the material and social precarity experienced by urban waste workers.

Over the last decade, solid waste has become a flagship area of interest and intervention for urban and national governance in India. Since the initiation of the Swachh Bharat Abhiyan (Clean India Mission) in 2014 by the Hindu nationalist Bharatiya Janata Party government at the central level, a policy environment has emerged which simultaneously aims to catalyse both cleanliness entrepreneurship and digital innovation. A series of governance initiatives, missions and funds have since followed the Swachh Bharat Abhiyan with the aim of transforming the economic, governance and infrastructural efficiency of cities, often through the development of technological solutions (Luthra et al. 2023; Parkar et al. 2023). A recent proliferation of tech start-ups and firms is leveraging machinery, sensors, digital platforms, biometric software and big data analytics in order to improve the collection, transportation and processing of solid waste across India. In this context, waste is being framed as an economic resource and an area of 'untapped' entrepreneurship and potential profit (Gidwani 2013; Luthra 2018).

Although solid waste management is the legal responsibility of urban local bodies in India, waste collection and recycling services have long been provided by a combination of municipal employees and self-employed or informal-sector workers. In Ahmedabad, unsegregated household waste was deposited by residents into large metal community waste bins positioned throughout the city, and municipal workers collected the bins in trucks to transport them to waste transfer stations and/or the local Pirana dumpsite. In this system, bin pick-ups were not scheduled or predictable and waste was often overflowing out of the bins, which led community members and street sweepers to set fires in the bins to reduce the volume of waste. The system was supplemented by informal recyclers (or waste pickers) who collected recyclable materials with any recoverable value (plastics, cardboard, paper, metal, hair, etc.) from discarded waste on roadsides, in community bins, in dumpsites or in open spaces to sell these materials and make a daily income. Often workers from low- or non-caste communities (Butt 2023), religious minority groups and interstate migrants (Kornberg 2019), including women (Dias and Samson 2016; Wittmer 2021), informal waste workers in India have long understood waste to be an essential economic resource. Yet, as municipal solid waste transforms into a 'new' market space dominated by tech start-ups and corporate capital, scholars and activists are noting a 'dispossession by platformisation' where a rise in platform-based business models is resulting in the replication and displacement of services long provided by established informal networks, activities and labour practices (Luthra and Monteith 2021).

In Ahmedabad, the city-wide uptake of a PPP model for urban solid waste collection in late 2016 led to the outsourcing of door-to-door waste collection services to private contractors using mechanised systems for hauling waste. This shift toward mechanised waste collection initially led to the masculinisation of waste labour in the city (Wittmer 2023a). However, seasonal migrants from Scheduled Tribe[1] communities (colloquially known as 'tribal' communities) from the borderlands of Dahod (eastern Gujarat) and Jhabua (western Madhya Pradesh) have since entered the door-to-door waste collection workforce in large numbers (Singh and Mane 2023). Studies on seasonal migration and the trans-local lives of labour migrants in the tribal belt area of Gujarat, Rajasthan and Madhya Pradesh highlight the ways in which migration serves as a means to cope with below-subsistence agriculture and debt, and as insurance against increasing environmental risks/failures in small-scale agriculture (Coffey et al. 2015; Desai 2017; Deshingkar 2008; Mosse et al. 2005). Workers from tribal communities in specific regions tend to be linked to particular types of labour in the city – for example, seasonal migrants from the tribal belt of Gujarat, Rajasthan and Madhya Pradesh constitute the majority of the construction workforce in Ahmedabad, where entire households migrate to the city together (Desai and Sanghvi 2018). However, amidst shifts in waste collection practices accompanying the roll-out of the PPP model and COVID-19 pandemic lockdowns halting construction work across the state, migrant workers from the tribal belt area have come to dominate Ahmedabad's current 'municipal' waste collection workforce.

With a focus on the emerging migrant workforce taking up waste collection work in Ahmedabad, this chapter explores the tensions underlying the processes through which precarious labour is made simultaneously visible and invisible in the informational periphery. While migrant workers are made visible through digital surveillance mechanisms and are monitored (digitally and in person) in 'real time', they are also rendered invisible in official systems because of employment models that exclude them from various social and labour entitlements. These informational gaps are key in producing the paradoxes of the informational periphery, as workers continue to live in poor infrastructural conditions of roadside migrant camps and lack access to various benefits and government schemes while they simultaneously provide infrastructural services to the city and produce data for municipalities and firms through the datafication and surveillance of their activities. Through the emerging transformations in the governance and management of solid waste in Ahmedabad, I argue that waste and waste labour are not only an economic resource for capital accumulation but are also an emerging informational resource, providing data for state and corporate actors alike in the governance of 'smart' cities. Further, as more and more data are produced to enable the circulation and management of waste materials and workers, urban solid waste becomes an important site through which urban informational peripheries are produced and where various material, social and digital exclusions are materialised.

The insights on migrant work and the informational periphery that follow emerge from my ethnographic study on waste management and informal labour in Ahmedabad between 2016 and 2024.[2] However, the work highlighted in this chapter draws particularly upon site visits, observations and brief conversations with 11 migrant workers engaged in waste collection work in two worker settlements/camps in Ahmedabad in 2023 and 2024. Many of my reflections about working conditions and labour entitlements emerge from my longer-term observations of informal waste work in the city and the reflections of local activists and labour organisers from the Centre for Labour Research and Action (CLRA) and the Bandhkaam Mazdoor Sangathan. Through this research, I continue to support these groups in their efforts to reach and advocate for the rights of waste workers in the city. With that said, my direct verbal interactions with workers themselves were quite brief and mediated by activist colleagues to limit the amount of time we spent in the worker settlements and avoid drawing attention from the contractors. Emerging from Ahmedabad, this research also includes evidence from media, policy documents and reports from civil society organisations and municipalities to inform a wider perspective on datafication trends and the production of the informational periphery through waste management labour in India.

The informational periphery and the datafication of infrastructural labour

Datafication is a contemporary process whereby human life is quantified, captured, analysed and circulated through digital information (Mejias and Couldry 2019). The process of transforming human experiences into digital data is now viewed as best practice in governing systems and societies, where the production of large datasets and the use of algorithmic analyses are understood to be informing decision making, increasing efficiency and improving urban life. However, scholars of critical data and surveillance studies highlight several areas of concern regarding the power asymmetries that are further entrenched and/or newly created through the datafication of life in highly uneven societies (Wittmer et al. 2025; Ash et al. 2018; Dalton et al. 2016; Shaw and Graham 2017). Of particular concern are the ways in which state and non-state actors alike are able to accrue value from the human experiences comprising digital information and the murky ethical contentions around the ownership of data which may or may not be taken with the consent or control of the urban populations who produce them (Dalton et al. 2016; Graham and Dittus 2022; Taylor and Mukiri-Smith 2021). As human life is increasingly datafied and, in turn, capitalised upon in this digital age, Ayona Datta (2024) suggests that the territorial notion of the periphery can be productively engaged through an informational lens to encompass the entanglements of digital and analogue forms of data with the politics and social relations which enable (and disable) its circulation and uses (Shaw and Graham 2017).

Peripheral urbanisation

Scholars engaging with the urban periphery note that the periphery is not merely a place or geographical location, but a 'conceptual and territorial threshold' which may be located at the edge of metropolitan cities (Gururani et al. 2021, 2). Theories of peripheral urbanisation note the heterogeneity of peripheral spaces and ways of being in the urban margins over time, highlighting that peripheries are necessarily incomplete, are always being transformed, and are 'undoubtedly about inequality' (Caldeira 2017, 9). Yet while Caldeira argues that residents are the primary producers of space in the urban periphery (that is, through the autoconstruction of settlements), Ashima Sood's (2021) work on India's speculative frontier highlights that India's peri-urban spaces are 'hostile' spaces for working-class and migrant populations, undercutting aspirations for 'occupancy urbanisms' as the interests of property groups and modes of privatised governance dominate development in this context (Sood 2021, 91; see also Benjamin 2008).

Anita Say Chan's (2013) conceptualisation of 'networked peripheries' resonates across this divergence. She argues that networked peripheries are characterised by fragmented relations, modes of governance and forms of infrastructural violence, which touch down in various ways in peripheral spaces over time. Through a networked informational lens, populations comprising the 'urban majority' may engage in infrastructural labours of autoconstruction, maintenance and repair in their settlements, but might also be increasingly displaceable and 'made the objects through which states and other regulatory apparatuses demonstrate their capacity to control, measure, and ... contain' in the context of fragmented governance and splintered relations with the state (Simone and Rao 2021, 55). The situating of peripheralised spaces and experiences within networks and in relation to fragmented institutions and infrastructures serves to reveal the multiplicity of social, material and digital entanglements of marginality in different contexts. This political project of situating the periphery in relation to power is instructive for investigating the material production of the informational periphery as spaces which are characterised simultaneously by infrastructural exclusions and informational extraction.

In the context of India's digitalising state, Datta theorises the informational periphery as 'spaces of informational exclusion and fragmentation in the margins of political-informational rights to territory' (Datta 2023, 151). She argues that social and material exclusions produced in the informational periphery are 'marked by both geographic and informational distance' from the state and that it is often composed of populations and places which are uncountable and unmappable (Datta 2023). In other words, the informational periphery consists of territories and populations which experience informational and digital precarity, invisibility and exclusion, but may also be simultaneously targeted for informational extraction by both state and non-state actors alike. Migrant workers coming to Ahmedabad for waste collection work are a fitting example of the interweaving of territorial and informational processes of peripheralisation. Here, the 'smart' digital transformation of the city leads to various aspects of workers' lives as infrastructural workers being made visible as data through surveillance mechanisms while simultaneously being territorially and socially excluded through labour sub-contracting in PPP models.

Infrastructural exclusions and labour

Amidst an 'infrastructural turn' in the social sciences (see Amin 2014; Harvey and Knox 2012; Kooy and Bakker 2008; Larkin 2013), geographies of infrastructure situated in South Asia have contributed a focus on the fragmented and inequitable socio-material relations that are continually (re)produced through the politics of distribution and material conditions of water, sanitation, housing, transportation and electricity infrastructures in the social and territorial peripheries of cities in the majority world (Anand 2017; Castán Broto and Sudhira 2019; Sultana 2020). However, recent work on infrastructure has called for a more explicit focus on labour in order to understand the ways through which 'infrastructures enrol, shape, and organise human labour' and how human labour sustains infrastructures and reproduces socio-material relations (Stokes and De Coss-Corzo 2023, 437; see also Strauss 2020). Theorisations of bodies as/and infrastructure frequently invoke AbdouMaliq Simone's notion of 'people as infrastructure' (Simone 2004, 407) to investigate the everyday and relational practices involved in accessing, repairing and even performing infrastructures in the majority world (Andueza et al. 2021; Desai et al. 2015; Ramakrishnan et al. 2021; Simone 2021).

Empirical research on solid waste and recycling labour has been particularly useful for conceptualising embodied labour as the infrastructural foundation of cities. Working from the experiences of communities engaged in recycling work in India, Vinay Gidwani and collaborators have theorised infrastructural labour as being essential to the production and maintenance of cities, services and capital, but often stigmatised and denied recognition (Gidwani 2015; Gidwani and Maringanti 2016; Corwin and Gidwani 2021). Feminist and anti-colonial ethnographies of waste labour have also provided important contributions toward understanding the embodied impacts of infrastructural services provided by cheap or free labour and their governance through regimes and practices of informality, gender, caste, race, class, religion and migration (Butt 2023; Kornberg 2019; Truelove and Ruszczyk 2022; Wittmer 2023a). Yet despite the essential nature of their work, waste workers are pushed to the social and geographical peripheries of cities. In this way, the people, practices, social relations and spaces that produce clean cities remain out of sight and are often denied recognition by state and civil society, unless to be blamed, condemned or cleaned up in moments of crisis (Gidwani and Maringanti 2016). In arguing for increased attention to labour in infrastructural studies, Stokes and De Coss-Corzo (2023) note that labour is an essential lens for understanding the ways in which human practices and infrastructure are coming together with the digital. As increasing amounts of data are now being produced through urban waste management services, solid waste and recycling labour has become an important site through which various material, social and digital exclusions are materialised.

Datafication and surveillance in the informational periphery

Critical studies of datafication and surveillance have focused significantly on relations of capitalism and colonialism within contemporary data regimes. Here, issues of labour appear in discussions of the political economy of data in terms of the role of digital labour in the production and maintenance of data infrastructures in increasing the circulation and accumulation of capital (Arefin and Prouse 2024; Attoh et al. 2019; Mahmoudi and Levenda 2016; Williams and Miceli 2023; Williams et al. 2022) and in the broader transformations of working conditions – including precaritisation, exploitation and shifts in social reproduction – in platformising economies around the world (Anwar and Graham 2020; Bissell 2020; Surie 2020).

A primary project of contemporary 'smart' digital urban governance is to 'see' through the digital – to make decisions and accrue value from big datasets as a 'temporal fix' to urban challenges (Datta 2023, 150, citing Scott 1999). In the context of solid waste and recycling processes, *seeing* may involve the datafication and surveillance of territories or geographic places, or circulations of bodies and equipment through geographic information systems and mapping technologies that were often previously made invisible in the context of global capital (Corwin and Gidwani 2021). These processes of datafication have recently been enacted in the lives of informal waste workers in India through drives to formalise their work. Attempts to formalise informal waste labour have necessitated both digital and analogue forms of information to record and categorise people, forms of labour, settlements, and other practices that were previously out of sight or uncontrolled by the state and/or which are considered illegal (Anantharaman 2019; Luthra and Monteith 2021; Wittmer 2023b). Investigations of urban informational peripheries necessitate a fundamental orientation toward grappling with relations of knowledge and power that are tied into practices of generating value from data extracted from the labour of disenfranchised urban populations.

This chapter proceeds by unpacking the data relations invoked through the infrastructural labour of garbage collection in producing informational peripheries in India. In theorising data colonialism, Couldry and Mejias (2019) define 'data relations' as a social process consisting of reconfiguring human relations to enable the extraction of data as an open resource for commodification, which they argue is a global process 'as expansive as historic colonialism's appropriation of land, resources, and bodies' (Couldry and Mejias 2019, 2). By centring infrastructural labour as a space in which data relations are rapidly unfolding, I aim to contribute to understandings of this contemporary moment of information extraction and commodification in urban India. The connections between information extraction, the visibility and consent of populations, and profit accumulation are of particular concern in India, which has been characterised as one of the 'most surveilled and data-hungry' systems in the world, where tech companies have increasing power and control in producing, owning and using extracted data (Chishti 2023, citing Petrosyan 2023).

By focusing on the data relations produced through the datafication of waste work, I respond to calls for centring labour relations produced through urban infrastructures and extend thinking on datafication beyond political economic and postcolonial critiques with grounded experiences of labour (and labour activism) that are being selectively surveilled and datafied by state and corporate actors. This work also aims to advance Datta's (2023; 2024) conceptualisation of the informational periphery by discussing how data relations produced in India's contemporary 'smart' waste management sector are leading to the informational and material precaritisation of urban migrant workers.

Public–private partnerships and modernising municipal waste collection in Ahmedabad

In recent years, two major events have directly resulted in widespread changes to the way solid waste is collected and managed in Ahmedabad. The first was the widespread privatisation of waste collection practices in Ahmedabad in early 2017 in response to revisions to the national solid waste management rules (MoEFCC 2016). The rules require door-to-door waste collection in cities. The Ahmedabad Municipal Corporation (AMC) responded to the rules by contracting out waste collection service delivery across all zones of the city to two private entities: Jigar Transport Company and Om Swachtha Corporation. This shift toward door-to-door collection and service delivery by private companies also entailed the removal of the community waste bins and tensions for informal workers who were losing access to waste in public space (Wittmer 2023a). Second, the COVID-19 pandemic lockdowns imposed in 2020 required informal workers to stay at home and the privatised collection services expanded and gained further legitimacy in the city. Further, because construction work had halted during the lockdowns, tribal migrant workers in the city began to take up the work of collecting waste on contract with Jigar Transport Company and Om Swachtha Corporation, and the role of waste collection in Ahmedabad is now dominated by migrant workers – predominantly from the Bhil tribe from Jhabua (Madhya Pradesh) and Dahod (Gujarat).

When I returned to Ahmedabad in 2022, I immediately noticed the shifts in waste collection practices: vehicles were being operated by family groups – men, women and children – involved not only in collecting and transporting waste, but also in separating out the recyclables to sell and earn additional income. I began to speak with local labour organisers, activists and scholars about this shift and they confirmed my observations, emphasising that this is a new stream of work for migrants – it is only in the last five years that tribal workers have been involved in waste collection. My interlocutors attributed this shift to both the increasing demand for workers, particularly for 'cheap labour' in the privatised door-to-door system, and the COVID-19 lockdowns which halted the usual streams of income for migrants from the tribal belt. In conversations with migrant workers in informal worker settlements, I asked how long they had been engaged in this work, and most indicated that they had been involved for less than four years; only one man responded that he had been working as a driver for over 10 years.

An additional shift in Ahmedabad's municipal waste collection practices pertains to the use of various digital technologies in tracking waste pick-ups and the movement of workers. This trend very much aligned with the digitalising state's push toward creating 'smart cities' and with national competitions (the annual *swachh survekshan*) for urban cleanliness where points are given to municipalities for integrating ICT solutions into the production and analysis of data, real-time monitoring and reporting functions for managing waste and urban cleanliness. In Ahmedabad, a major function of scaling up a PPP model for waste collection in 2017 was to modernise these services through mechanisation (that is, the widespread presence of waste collection vehicles engaged in door-to-door collection) and to integrate ICT solutions into local waste management practices. Since then, Amnex Infotechnologies has served as the technology solutions provider for the AMC, equipping waste collection trucks with infrastructure such as global positioning systems (GPS), radio frequency identification (RFID) systems, and value and weight sensors in order to monitor the functioning and movement of vehicles.

Amnex has also developed a platform featuring two primary apps, Ecoskipper and Ecokeeper, which the AMC uses to monitor activities and generate and analyse data on waste management activities across the city. The platform's focus areas include waste collection and route optimisation, contractor management, missed collection notifications and incident alerts. It also offers a solid waste management operating interface, productivity analytics, and a mobile app through which citizens can report grievances. Additionally, the app allows waste collection and sweeping staff to log their daily attendance data (Amnex n.d.). Other private companies in Ahmedabad have also developed platforms and use technology-based approaches to deliver more efficient waste collection services. For example, the for-profit company Nepra Resource Management Pvt Ltd has developed a company-wide platform with various apps tailored to specific functions, such as managing household and business recycling pick-up logistics, ensuring compliance with extended producer responsibility, and facilitating zero-waste events. Nepra also offers an app for waste pickers who sell to their trucks, enabling them to access their accounts in the field using facial recognition software. This allows them to view their full account history, including records of scrap sold and income owed or paid. This technology solution is intended to enhance informational access for individual workers by tracking their sales and creating a record of their interactions with the company. It also records the locations of their sales, which enables the company to generate a city-wide map of sale locations. ICT and technology solutions produced by firms and often facilitated through PPP models are becoming a key enabler of the emerging system of 'smart' solid waste management. Focusing on data relations in 'smart' waste collection practices is therefore increasingly important for understanding the relationship between waste, technology and labour, and for ensuring that informal workers are aware of and consulted about their informational rights.

Informational distance in the urban periphery

Under Ahmedabad's PPP model for waste collection, tribal migrant workers come to work for the municipality through several layers of sub-contracting. Under this model, staff hiring, management and salary payments are currently the responsibility of registered contractors, not the AMC. However, this model enables registered contractors who win the bidding process to further sub-contract staffing out to unregistered contractors or agencies, and the hiring itself is then passed down again to a '*mukadham*' or intermediary, connected with networks of seasonal migrants, who ultimately hires, supervises and controls workers' salaries and labour conditions. The *mukadham* is their only point of contact regarding employment contracts, salaries and working conditions. This multi-step distancing from the principal employer, the AMC, ensures that these workers have no formal affiliation with the municipality. As a result, most workers do not have access to fundamental labour entitlements such as unemployment insurance or benefits like the Employees' Provident Fund, which covers retirement savings, pensions, life insurance and savings contributions.

The workers I spoke with indicated that although they are usually hired in pairs, often as husband-and-wife teams, only the driver of each vehicle receives the monthly salary, which must then be shared among all family members working on the truck. The reported monthly salaries ranged from INR.10,000 to INR.20,000 for those driving larger trucks, while drivers of smaller trucks were paid based on the weight of the vehicle per collection round, earning between INR.6,000 and INR.12,000. All the workers I spoke with emphasised the importance of separating and selling recyclables daily, not only to supplement their salaries but as a vital source of income for daily expenses such as food and water. Scrap sales, ranging from INR.150 to INR.600 per day, were particularly crucial for meeting these needs. Those driving vehicles also noted that while the contractor provides the truck and fuel, any damage to the vehicle or equipment must be borne by the driver themselves. Furthermore, no workers I spoke with had ever been given any gloves, masks, boots, vests or other personal protective equipment for their work of collecting garbage.

The PPP system awards waste collection contracts to the lowest bidder, incentivising firms to bid low to secure contracts and then find ways to cut costs in service delivery. The labour organisers and activists I spoke with were clear in linking this system of contractualisation with the exploitation of tribal workers and profit imperatives. In other words, despite being highly visible workers in public space who are providing an essential municipal public service, migrant waste collectors in Ahmedabad remain informationally invisible within municipal systems, thereby being excluded from the rights that their labour should grant them. The absence of insurance and security for waste collection workers underscores the bureaucratic gaps that exist within informational systems, where workers cannot gather the documents required to prove their identity or residence in the city to access various benefits or welfare systems without forfeiting their status and documents from their village (also see Desai and Sanghvi 2018). As a result, migrant workers exist informally in the city in settlement camps located near waste transfer stations or within spaces provided by contractors for parking collection trucks. Despite national programmes like 'One Nation One Ration Card', the realities of stigmatised workers navigating access to rations while living in labour camps at the territorial periphery of the city prevent them from gaining full access to such schemes. Workers and activists alike noted that either no rations were ever accessed or, if they were able to access them, the rations were given in smaller amounts than promised or were not foods they would normally cook with or consume.

Migrant waste workers are also subject to regular displacement and threats to their settlements at the periphery of the city. In early 2023, I spoke with workers living in an open area along a roadside, just outside a wall enclosing a waste transfer station and an adjacent empty plot. The union activist accompanying me was shocked, asking what had happened to the five hundred or so huts that had previously stood within the walled plot next to the waste transfer compound just a few weeks earlier. The workers explained that the municipality had demolished their huts (*kutcha* tents/huts) and forced them to leave that space. The families spread out around the block, with many occupying the roadside along the wall of the waste transfer station. The workers speculated that the municipality wanted to build 'another' building – like the new 'luxury shops and offices' building nearing completion across the road, with signs advertising a second building 'coming soon' by the same development company in the adjacent lot. The workers we spoke to said this was the third time in two years that their huts had suddenly been taken down by the municipality. Indeed, reviewing satellite image histories of the space, I observed the spatial shifts in the migrant camp's composition, size and location around the empty lots and roadsides in this area in response to the various development projects emerging around the site. In this context, migrant tribal waste workers are continually displaced and made to shift their temporary dwellings in response to various projects as peripheral spaces suddenly become valuable – such as when a new highway connected this space to the city's broader transit networks. These settlements comprise the 'essential surrounds' of cities – the places, people and activities which 'remain or become marginal to the official components of an urban economy' – as workers continually navigate states of flux, circulation and improvisation as they struggle to survive amid forces beyond their control across multiple sites (Simone 2022, 8).

By considering the infrastructural labour that tribal migrants perform and the conditions of their work and housing, one can observe the ways in which precarity is distributed in the periphery through institutions of wage labour and is socially produced by relations of caste, patronage and capitalism (Simone 2022; Strauss 2020). Given that migrant workers are hired through systems of sub-contracting, there are no official records of their employment as waste collectors in Ahmedabad; their labour thus serves as the 'the constitutive outsides [or peripheries] of capitalist enterprise' (Corwin and Gidwani 2021, 8). Here, migrant workers create value for others and ensure the continual functioning of urban systems until they are no longer needed and/or are made to disappear or otherwise shift. In Ahmedabad's territorial and informational periphery, unregistered migrant workers are constantly responding to the extractive rationalities of local institutions and regimes of capitalism and neoliberal governance which surround the provisioning of municipal waste collection services. However, despite being made invisible within state occupational information systems and the physical spaces of their urban settlements/camps, as informal workers and settlements have long experienced, these particular migrant waste collection workers are now simultaneously subjected to new forms of datafication and surveillance.

Informational surveillance and datafication in the urban periphery

Simultaneous to the informational distancing occurring between infrastructural work and the municipality as the principal employer, digital data on workers are continuously being produced through 'smart' technology solutions. As waste management has become an important area for ICT innovations (GPS, RFID, and closed-circuit television cameras on waste collection vehicles and positioned throughout the city) to generate more efficient, modern solutions to the problems posed by increasing urban waste production, waste workers have become the site through which these solutions are enabled and produced.

When I asked the workers if they had noticed any changes to their work in the last year, two of the drivers of larger vehicles answered that GPS trackers had been installed on all the trucks. When I enquired what the GPS was for, they responded, 'to track us, to know where we are going' and 'to know whether we are resting or doing work'. In official documents and reports on waste management planning, these technological and surveillance solutions are said to enable municipal officials and supervisors to track the 'progress' of waste collection in the city's Integrated Command and Control Centre (a hallmark of India's Smart City Mission, present in cities across the country). These sensors provide feedback to the authorities via the Amnex platform and its integrated Ecoskipper and Ecokeeper apps. Ecoskipper serves as a 'centralised platform' to facilitate the 'end-to-end management of solid waste collection' and 'complete the collection network from administrator, to vehicle, to citizen' (by planning/tracking vehicle pick-up points and routes). Meanwhile, Ecokeeper enables supervisors and officials to monitor sweeping and sanitation staff attendance on location in real time using GPS sensors, and 'empowers the average citizen with avenues for active participation', such as sharing 'grievances and feedback' on the platform (Amnex n.d.). According to Amnex, Ecokeeper also provides supervisors with registration and attendance data on on-location sweeping and waste collection staff through biometric software.

An ICT engineer consulting on developing software for a different 'smart city' development corporation in India explained that absenteeism is a significant issue among sanitation workers. In response, this municipal corporation has introduced smartwatches – referred to as 'human efficiency tracking systems' (Inzamam and Qadri 2022). The engineer emphasised the role of partnerships between municipal corporations and tech companies, such as the Indian firm OneMap, in developing these smart innovations, which provide much-needed discipline and enhance surveillance in relation to workers. By tracking worker movements via GPS-enabled watches and systems installed on trucks, municipal corporations are using technology interventions to assert that they are increasing the efficiency of the waste collection system and the overall cleanliness of the city.

GPS-enabled smartwatch devices have been introduced in a number of Indian cities, including Chandigarh, Indore, Nagpur, Noida and Pune, where municipal sweepers are employed to sweep roadsides and clean up garbage. The devices are described as being 'fitted with GPS trackers, a microphone, a SIM embedded for calling workers on the device, and a camera so the workers can send photos to their supervisors as proof of attendance' (Inzamam and Qadri 2022). When I asked the software developer about the uptake of the watches, he explained that while there had been some initial issues around onboarding workers (such as workers losing watches, refusing to wear them or burning them), most had now come round to the watches because the municipality had taken up a policy of 'no data, no pay' and withheld salary payments from those who were not adhering to the programme. Ahmedabad is not yet using smartwatches, and I asked the consultant how waste collection workers in Ahmedabad living in *kutcha* settlements, which often have no electricity, would charge the devices. He conceded that it is not yet possible to use these devices in all cities. In cities like Mumbai and Ahmedabad, workers check in by taking a selfie with their supervisor on the supervisor's phone and they submit this proof of attendance on the platform.

There is no clear public information available regarding what happens to the worker data that is produced through GPS tracking and attendance records in Ahmedabad, beyond their monitoring at the local Integrated Command and Control Centre. While worker information data is currently accessible only to supervisors and municipal authorities, other cities with similar digital initiatives that have partnered with OneMap and are using smartwatches (such as the Greater Noida Authority) have not only started to post 'complete, real-time information' for tracking the city's waste collection using GPS data, but have also reportedly started uploading on their websites 'the details of all sanitation workers, supervisors, and private agencies engaged in cleaning the residential sectors and villages across the city' (Sinha 2021). These applications of real-time data, presumably produced in the name of transparency, raise important ethical questions about the extraction and ownership of data and who these processes are creating 'transparency' for. They also point to the potential for 'surveillance creep', where a 'leading model' technology such as the smartwatch and digital and algorithmic worker-tracking platforms can move from place to place without clear ethical or context-specific guardrails.

Waste collection labour as an economic and informational resource

This chapter has explored the data relations that shape and are shaped by the modernisation of urban solid waste management systems in India through the privatisation, datafication, surveillance and platformisation of solid waste management labour practices. The case of tribal solid waste collection workers in Ahmedabad highlights the informational tensions that define workers' experiences of social, material and digital precarity in the spaces between their informational visibility (surveillance) and invisibility (exclusion). While contracted migrant workers are made visible through digital surveillance mechanisms and their data is algorithmically and manually monitored in 'real time', they are also rendered invisible in official systems which would enable them to access housing, daily resources and urban space. While information and data about labour is produced as a resource for the state, employers and civil society, this information also precaritises workers as bureaucratic and scalar mismatches clash with the requirements of urban service delivery systems which rely on seasonal labour migration. These informational gaps are key in producing the paradoxes of the informational periphery, where workers continue to live in poor infrastructural conditions, such as roadside migrant camps, and lack access to various benefits and government schemes. At the same time, they endure significant health risks and contribute to the city's infrastructure while their everyday activities are datafied and surveilled, producing valuable information for municipalities and private companies.

While India's smart cities are catalysing their commitments to technology solutions and the national cleanliness mission through the application of GPS sensors, cameras, apps/platforms, smartwatches ('human efficiency tracking systems') and biometric software, this chapter illustrates how these systems of datafication and surveillance are producing smartness through the exploitation of worker bodies and experiences. The immense amount of data generated about precarious work for state and private actors in this sector raises fundamental questions about the power asymmetries involved in the extraction, interpretation and ownership of these forms of digital information. Anti-colonial critiques of big data and surveillance offer insights into the ways in which people are 'dispossessed and alienated from the data they generate' and from the right to control that data (Thatcher et al. 2016, 998; see also Dalton et al. 2016; Nost and Goldstein 2022). The real-time tracking and production of data on waste collection services and the recording of worker activities – captured through GPS trackers, watches and cameras – is particularly concerning in the context of seasonal migrant labour in solid waste management. These workers rarely have access to smartphone technologies, the digital literacy to use apps, or the knowledge or pathways through which to hold the state accountable because of the current PPP model of sub-contractualisation.

The case of tribal seasonal migrant workers in Ahmedabad's PPP waste collection system reveals the simultaneous and selective practices of surveilling workers' bodies and activities while rendering them unrecorded and fungible (indistinguishable, replaceable manual labour). These concurrent processes can be understood as keeping workers in the city's informational periphery: they are being observed by the state, corporate actors and even civil society through sensors, cameras and digital platforms, while simultaneously being excluded from employment records and labour entitlements. Their experiences are datafied in a networked periphery, which is also a network of marginality (Chan 2013), as they are simultaneously and selectively seen and unseen by fragments of institutions, programmes, technologies and infrastructures across a digitalising public–private interface. The notion of the informational periphery is essential here for grounding interrogations into the social, material and digital exclusions that workers experience. For example, looking at the informational periphery produced through waste labour involves considering the infrastructural realities of migrant worker camps, as well as workers' lack of access to protective equipment, housing and urban entitlements, and the selective processes involved in surveillance, data production and information sharing on workers by municipal actors and contracting firms.

Notes

1 The Scheduled Tribes have been an official bureaucratic category in the Constitution of India since the country's independence. This category recognises Indigenous identity and grants members of designated communities access to affirmative action programmes. Despite this, many individuals classified under the Scheduled Castes and Scheduled Tribes (SC/ST) categories continue to face persistent and multidimensional deprivation and injustice. While many Indigenous groups in India identify as 'Adivasi' (meaning 'earliest inhabitants') as a political term, in the northeastern and Himalayan states, the term 'tribal' is more commonly used by Scheduled Tribe groups (Cháirez-Garza et al. 2022; Ranganathan 2022).

2 The initial study (2016–18) was based on a survey of 401 women informal recyclers across 10 areas of Ahmedabad, in-depth interviews with 45 women workers from the sample, follow-up interviews with 36 participants, and focus group discussions with 12 groups of participants in their homes and worksites. The study also relied on interviews with local scholars and colleagues in advocacy and labour organisations and an ongoing discourse analysis of policy and media (see Wittmer and Qureshi 2023; Wittmer 2021, 2023a, 2023b).

References

Amin, A. 2014. Lively infrastructure. *Theory, Culture & Society* 31(8): 137–61. https://doi.org/10.1177/0263276414548490.

Amnex. (n.d.). From pick up to disposal: Waste management done waste-free. Accessed February 2025. https://www.amnex.com/solution.html?type=Solid-Waste-Management.

Anand, N. 2017. *Hydraulic City: Water and the infrastructures of citizenship in Mumbai*. Durham, NC: Duke University Press.

Anantharaman, H. 2019. Neo-liberalising inclusion? Waste picking, data activism, and the state. *Economic & Political Weekly* 54(47). Accessed February 2025. https://www.epw.in/journal/2019/47/review-urban-affairs/neo-liberalising-inclusion.html.

Andueza, L., Davies, A., Loftus, A. and Schling, H. 2021. The body as infrastructure. *Environment and Planning E: Nature and space* 4(3): 799–817. https://doi.org/10.1177/2514848620937231.

Anwar, M. A. and Graham, M. 2020. Hidden transcripts of the gig economy: Labour agency and the new art of resistance among African gig workers. *Environment and Planning A: Economy and space* 52(7): 1269–91. https://doi.org/10.1177/0308518X19894584.

Arefin, M. R. and Prouse, C. 2024. Urban political ecologies of sewage surveillance: Creating vital and valuable public health data from wastewater. *Transactions of the Institute of British Geographers*, 7 December. https://doi.org/10.1111/tran.12732.

Ash, J., Kitchin, R. and Leszczynski, A. 2018. Digital turn, digital geographies? *Progress in Human Geography* 42(1): 25–43. https://doi.org/10.1177/0309132516664800.

Attoh, K., Wells, K. and Cullen, D. 2019. 'We're building their data': Labor, alienation, and idiocy in the smart city. *Environment and Planning D: Society and space* 37(6): 1007–24. https://doi.org/10.1177/0263775819856626.

Benjamin, S. 2008. Occupancy urbanism: Radicalizing politics and economy beyond policy and programs. *International Journal of Urban and Regional Research* 32(3): 719–29. https://doi.org/10.1111/j.1468-2427.2008.00809.x.

Bissell, D. 2020. Affective platform urbanism: Changing habits of digital on-demand consumption. *Geoforum* 115: 102–10. https://doi.org/10.1016/j.geoforum.2020.06.026.

Butt, W. 2023. *Life Beyond Waste: Work and infrastructure in urban Pakistan*. Redwood City, CA: Stanford University Press.

Caldeira, T. P. R. 2017. Peripheral urbanization: Autoconstruction, transversal logics, and politics in cities of the global south. *Environment and Planning D: Society and space* 35(1): 3–20. https://doi.org/10.1177/0263775816658479.

Castán Broto, V. and Sudhira, H. S. 2019. Engineering modernity: Water, electricity and the infrastructure landscapes of Bangalore, India. *Urban Studies* 56(11): 2261–79. https://doi.org/10.1177/0042098018815600.

Cháirez-Garza, J. F., Denzin Gergan, M., Ranganathan, M. and Vasudevan, P. 2022. Introduction to the special issue: Rethinking difference in India through racialization. *Ethnic and Racial Studies* 45(2): 193–215. https://doi.org/10.1080/01419870.2021.1977368.

Chan, A. S. 2013. *Networking Peripheries: Technological futures and the myth of digital universalism.* Cambridge, MA: MIT Press.

Chishti, S. 2023. G20: India is now the Vishwaguru of digital authoritarianism. *The Wire*, 17 August. Accessed February 2025. https://thewire.in/rights/g20-india-is-now-the-vishwaguru-of-digital-authoritarianism.

Coffey, D., Papp, J. and Spears, D. 2015. Short-term labor migration from rural North India: Evidence from new survey data. *Population Research and Policy Review* 34(3): 361–80. https://doi.org/10.1007/s11113-014-9349-2.

Corwin, J. E. and Gidwani, V. 2021. Repair work as care: On maintaining the planet in the capitalocene. *Antipode*, 1–20. https://doi.org/10.1111/anti.12791.

Couldry, N. and Mejias, U. 2019. Data colonialism: Rethinking big data's relation to the contemporary subject. *Television and New Media* 20(4): 336–49. https://doi.org/10.1177/1527476418796632.

Dalton, C. M., Taylor, L. and Thatcher, J. 2016. Critical data studies: A dialog on data and space. *Big Data & Society* 3(1). https://doi.org/10.1177/2053951716648346.

Datta, A. 2023. The digitalising state: Governing digitalisation-as-urbanisation in the global south. *Progress in Human Geography* 47(1): 141–59. https://doi.org/10.1177/03091325221141798.

Datta, A. 2024. The informational periphery: Territory, logistics and people in the margins of a digital age. *Asian Geographer* 41(2): 125–42. https://doi.org/10.1080/10225706.2023.2253233.

Desai, R. 2017. *Entitlements of Seasonal Migrant Construction Workers to Housing, Basic Services and Social Infrastructure in Gujarat's Cities: A background policy paper*. Centre for Urban Equity, CEPT University, Ahmedabad. Accessed February 2025. https://cept.ac.in/UserFiles/File/CUE/Working%20Papers/Revised%20New/WP%2035%20revised%20-%20Entitlements%20of%20Seasonal%20Migrant%20Construction%20Workers.pdf.

Desai, R., McFarlane, C. and Graham, S. 2015. The politics of open defecation: Informality, body, and infrastructure in Mumbai. *Antipode* 47(1): 98–120. https://doi.org/10.1111/anti.12117.

Desai, R. and Sanghvi, S. 2018. *Case Study: Migrant construction workers' housing in Ahmedabad.* Centre for Urban Equity, CEPT University, Ahmedabad. Accessed February 2025. https://mdl.donau-uni.ac.at/binucom/pluginfile.php/405/mod_page/content/38/P4_WP2.5_Case_Study_3_Migrant_Constr_CEPT.pdf.

Deshingkar, P. 2008. Circular internal migration and development in India. In *Migration and Development Within and Across Borders: Research and policy perspectives on internal and international migration*, edited by J. DeWind and J. Holdaway, 163–90. Geneva: International Organization for Migration.

Dias, S. M. and Samson, M. 2016. *Informal Economy Monitoring Study Sector Report: Waste pickers*. WIEGO. Accessed February 2025. https://www.wiego.org/sites/default/files/publications/files/Dias-Samson-IEMS-Waste-Picker-Sector-Report.pdf.

Gidwani, V. 2013. Six theses on waste, value, and commons. *Social & Cultural Geography* 14(7): 773–83. https://doi.org/10.1080/14649365.2013.800222.

Gidwani, V. 2015. The work of waste: Inside India's infra-economy. *Transactions of the Institute of British Geographers* 40(4): 575–95. https://doi.wiley.com/10.1111/tran.12094.

Gidwani, V. and Maringanti, A. 2016. The waste-value dialectic: Lumpen urbanization in contemporary India. *Comparative Studies of South Asia, Africa and the Middle East* 36(1): 112–33. https://doi. 10.1215/1089201x-3482159.

Graham, M. and Dittus, M. 2022. *Geographies of Digital Exclusion: Data and inequalities*. London: Pluto Press.

Gururani, S., Kennedy, L. and Sood, A. (eds). 2021. *Engaging the Urban from the Periphery. South Asia Multidisciplinary Academic Journal* 26 (March). https://doi.org/10.4000/samaj.7131.

Harvey, P. and Knox, H. 2012. The enchantments of infrastructure. *Mobilities* 7(4): 521–36. https://doi.10.1080/17450101.2012.718935.

Inzamam, Q. and Qadri, H. 2022. In India, digital snooping on sanitation workers. *UNDARK*. Accessed February 2025. https://undark.org/2022/05/02/in-india-digital-snooping-on-sanitation-workers.

Kooy, M. and Bakker, K. 2008. Technologies of government: Constituting subjectivities, spaces, and infrastructures in colonial and contemporary Jakarta. *International Journal of Urban and Regional Research* 32(2): 375–91. https://doi.org/10.1111/j.1468-2427.2008.00791.x.

Kornberg, D. 2019. From Balmikis to Bengalis: The 'casteification' of Muslims in Delhi's informal garbage economy. *Economic & Political Weekly* 54(47). Accessed February 2025. https://www.epw.in/journal/2019/47/review-urban-affairs/balmikis-bengalis.html.

Larkin, B. 2013. The politics and poetics of infrastructure. *Annual Review of Anthropology* 42: 327–43. https://doi.org/10.1146/annurev-anthro-092412-155522.

Luthra, A. 2018. Municipalization for privatization's sake. *Society and Business Review* 14(2): 135–54. https://doi.10.1108/sbr-11-2017-0102.

Luthra, A. and Monteith, W. 2021. Of market vendors and waste collectors: Labour, informality, and aesthetics in the era of world-class city making. *Antipode* 55: 1,068–88. https://doi.org/10.1111/anti.12784.

Luthra, A., Lowe, J. and Ochoa Berkley, K. 2023. Dispossession by platformization: The rise of on-demand recycling in urban India. *Geoforum* 141. https://doi.org/10.1016/j.geoforum.2023.103731.

Mahmoudi, D. and Levenda, A. 2016. Beyond the screen: Uneven geographies, digital labour, and the city of cognitive-cultural capitalism. *TripleC* 14(1): 99–120. https://doi.org/10.31269/triplec.v14i1.699.

Mejias, U. A. and Couldry, N. 2019. Datafication. *Internet Policy Review* 8(4): 1–10. https://doi.10.14763/2019.4.1428.

MoEFCC. 2016. Solid waste management rules 2016. New Delhi: Ministry of Environment, Forest and Climate Change. Accessed February 2025. https://cpcb.nic.in/rules-2.

Mosse, D., Gupta, S. and Shah, V. 2005. On the margins in the city: Adivasi seasonal migration in Western India. *Economic & Political Weekly* 40(28).

Nost, E. and Goldstein, J. E. 2022. A political ecology of data. *Environment and Planning E: Nature and space* 5(1): 3–17. https://doi.10.1177/25148486211043503.

Parkar, K., Zerah, M. H. and Mittal, G. 2023. Platformisation, infrastructuring, and datafication: Regional variations in the digitalisation of Indian cities. *Economic & Political Weekly* 58(14). Accessed February 2025. https://www.epw.in/journal/2023/14/special-articles/platformisation-infrastructuring-and-datafication.html.

Petrosyan, A. 2023. Most surveilled countries worldwide in 2022, by number of people affected. Statista. Accessed February 2025. https://www.statista.com/statistics/1290708/top-surveilled-countries-worldwide.

Ramakrishnan, K., O'Reilly, K. and Budds, J. 2021. Between decay and repair: Embodied experiences of infrastructure's materiality. *Environment and Planning E: Nature and space* 4(3): 669–73. https://doi.org/10.1177/2514848620980597.

Ranganathan, M. 2022. Caste, racialization, and the making of environmental unfreedoms in urban India. *Ethnic and Racial Studies* 45(2): 257–77. https://doi.10.1080/01419870.2021.1933121.

Scott, J. C. 1999. *Seeing Like a State: How certain schemes to improve the human condition have failed*. New Haven, CT: Yale University Press.

Shaw, J. and Graham, M. 2017. An informational right to the city? Code, content, control, and the urbanization of information. *Antipode* 49(4): 907–27. https://doi.org/10.1111/anti.12312.

Simone, A. 2004. People as infrastructure: Intersecting fragments in Johannesburg. *Public Culture* 16(3): 407–29. https://doi.org/10.1215/08992363-16-3-407.

Simone, A. 2021. Ritornello: 'People as infrastructure'. *Urban Geography* 42(9): 1,341–48. https://doi.org/10.1080/02723638.2021.1894397.

Simone, A. 2022. *The Surrounds: Urban life within and beyond capture*. Durham, NC: Duke University Press.

Simone, A. and Rao, V. 2021. Counting the uncountable: Revisiting urban majorities. *Public Culture* 33(2): 151–60. https://doi.org/10.1215/08992363-8917150.

Singh, A. and Mane, P. 2023. Navigating through waste: A study on the condition of migrant tribal families engaged in door-to-door waste collection work. Centre for Labour Research and Action. Accessed February 2025. https://clra.in/wp-content/uploads/2024/06/f6c774eb-e7d6-4d21-ac84-8d3aaae9b6e2.pdf.

Sinha, M. 2021. Irregular door-to-door collection could be a thing of the past: You can now track trash trucks in Gr Noida. *The Times of India*, 21 December. Accessed February 2025. https://timesofindia.indiatimes.com/city/noida/irregular-door-to-door-collection-could-be-a-thing-of-the-past-you-can-now-track-trash-trucks-in-gr-noida/articleshow/88298510.cms.

Sood, A. 2021. The speculative frontier: Real estate, governance and occupancy on the metropolitan periphery. *South Asia Multidisciplinary Academic Journal* 26: 71–99. https://doi.org/10.4000/samaj.7204.

Stokes, K. and De Coss-Corzo, A. 2023. Doing the work: Locating labour in infrastructural geography. *Progress in Human Geography* 47(3): 427–46. https://doi.org/10.1177/03091325231174186.

Strauss, K. 2020. Labour geography III: Precarity, racial capitalisms and infrastructure. *Progress in Human Geography* 44(6): 1,212–24. https://doi.10.1177/0309132519895308.

Sultana, F. 2020. Embodied intersectionalities of urban citizenship: Water, infrastructure, and gender in the Global South. *Annals of the American Association of Geographers* 110(5): 1407–24. https://doi.org/10.1080/24694452.2020.1715193.

Surie, A. 2020. On-demand platforms and pricing: How platforms can impact the informal urban economy, evidence from Bengaluru, India. *Work Organisation, Labour and Globalisation* 14(1): 83–100. https://doi.org/10.13169/workorgalaboglob.14.1.0083.

Taylor, L. and Mukiri-Smith, H. 2021. Human rights, technology and poverty. In *Research Handbook on Human Rights and Poverty*, edited by M. F. Davis, M. Kjaerum and A. Lyons, 535–49. Cheltenham: Edward Elgar.

Thatcher, J., O'Sullivan, D. and Mahmoudi, D. 2016. Data colonialism through accumulation by dispossession: New metaphors for daily data. *Environment and Planning D: Society and space* 34(6): 990–1006. https://doi.org/10.1177/0263775816633195.

Truelove, Y. and Ruszczyk, H. A. 2022. Bodies as urban infrastructure: Gender, intimate infrastructures and slow infrastructural violence. *Political Geography* 92. https://doi.org/10.1016/j.polgeo.2021.102492.

Williams, A. and Miceli, M. 2023. Data work and its layers of (in)visibility. *Just Tech*, 6 September. https://doi.org/10.35650/JT.3060.d.2023.

Williams, A., Miceli, M. and Gebru, T. 2022. The exploited labor behind artificial intelligence. *Noēma*, 23 October. Accessed February 2025. https://www.noemamag.com/the-exploited-labor-behind-artificial-intelligence.

Wittmer, J. 2021. 'We live and we do this work': Women waste pickers' experiences of wellbeing in Ahmedabad, India. *World Development* 140. https://doi.org/10.1016/j.worlddev.2020.105253.

Wittmer, J. 2023a. Dirty work in the clean city: An embodied urban political ecology of women informal recyclers' work in the 'clean city'. *Environment and Planning E: Nature & space* 6(2): 1343–65. https://doi.org/10.1177/25148486221102374.

Wittmer, J. 2023b. 'I salute them for their hard work and contribution': Inclusive urbanism and organizing women recyclers in Ahmedabad, India. *Urban Geography* 44(9): 1911–30. https://doi.org/10.1080/02723638.2023.2192560.

Wittmer, J., Prouse, C. and Arefin, M. R. 2025. Digitalizing sewage: The politics of producing, sharing, and operationalizing data from wastewater-based surveillance. *Environment and Planning C: Politics and space*, published online, 10 January. https://doi.org/10.1177/23996544241313454.

Wittmer, J. and Qureshi, M. 2023. Navigating the emotion-embodiment-language nexus in international research: Stories from a foreign researcher and local interpreter. *Emotion, Space and Society* 49. https://doi.org/10.1016/j.emospa.2023.100990.

4

Informational peripheries by design: land, layer and leverage in the Guadalajara Metropolitan Area

Juan Angel Demerutis Arenas, Ayona Datta and Jesus Flores

Visor Urbano (VU) is a digital platform aimed at improving urban planning and governance in cities, particularly focusing on the management of land and property data. It was established in 2016, when the municipality of Guadalajara became one of the five winners of the 'Mayors Challenge' convened by Bloomberg Philanthropies in the USA. Championed by the Mexican state as a technological solution during the urban reforms begun under mayor Enrique Alfaro's administration (2015–18), VU claimed to minimise human contact with public officials and thereby remove 'discretional' leverage in procedures related to urban planning. VU was seen as an innovative digital platform that would expand governance technologies (or GovTech) by providing open access information on land use to automate construction and business licensing in the municipality of Guadalajara.

In this chapter we argue that despite its aspirations of seamless, automated GovTech, VU produces informational peripheries by definition as it grafts itself onto earlier analogue and digital planning systems, thus inheriting the legacies of informational gaps and absences in the region. While the Guadalajara Metropolitan Area (GMA) is one of the pioneers of automated planning systems in Mexico, informational peripheries are designed into its digital land management platform via three contentious axes of power: land, layer and leverage. The GMA consists of nine municipalites: Zapopan, Tlajomulco de Zuñiga, Tonala, Tlaquepaque, El Salto, Juanacatlán, Ixtlahuacan de los Membrillos, Zapotlanejo and Guadalajara. Guadalajara, Zapopan and Tlajomulco are at the forefront of the digitalisation process.

First, we suggest that the dynamics between land and digital platforms are often fragmented, contested and ambiguous, resulting in multiple information gaps for governance and decision making. Land that has not been mapped or is outside the jurisdiction of the municipality does not make its way as information onto VU's digital platform, yet that does not deny its existence as a material place. Land here is juridical-technological by the fact that legal instruments which seek to liberalise land for the market also impose antagonistic (in/formal, il/legal) land categories, as in the case of indigenous land, that might lead to its exclusion from digital representation. This dualism determines how land itself becomes ambiguous in its representation as it is excluded or made invisible within digital platforms, and thus contributes to the limitations of the platform itself.

Second, digital platforms are grafted onto analogue technologies of statecraft, such as cadastral maps (maps of property ownership), which serve as the basis of future land management platforms. Here a single cadastral geographic information system (GIS) layer constitutes the foundation of VU's platform even though its accuracy remains a source of contestation across municipalities, which then challenges the interoperability, standardisation and maintenance of automated platforms. Thus, even though VU claims to present an accurate portrayal of land transactions in the region, the accuracy of territorial information on VU's automated platform is geographically ambiguous, asynchronous and fragmented, as cadastral layers are often fiercely protected and locked away in different municipalities in the GMA.

Third, while the decentralisation of the Mexican state (devolution of powers to sub-national and municipal governments) gave greater autonomy to municipal governments via constitutional amendments, it nonetheless made centralised digital platforms such as VU untenable. Despite VU's claim to reduce the 'discretion' of officials in manipulating planning processes, decentralisation has paradoxically increased municipal leverage regarding data sharing and information flows in the platform. This is particularly poignant when it comes to sharing cadastral layers, which represent a key revenue source for municipalities. Municipal leverage, then, emerges as a new form of decentralised power vested in the municipal body that has replaced earlier forms of discretion vested in the bodies of municipal officials.

Together, these three axes of power – land, layer and leverage – produce informational peripheries by design in a digital platform such as VU. The material-political-technological tensions inherent in these axes are built into the informational infrastructures onto which automated planning systems are grafted (Mattern 2017). The dynamics between land, layer and leverage suggest how informational peripheries are both geographies produced by the digital and geographies of the digital (Ash et al. 2018). Informational peripheries in the case of the GMA are therefore places that are inaccurately represented or invisible on the VU platform, as well as places that have historically been peripheral to metropolitan governance. These metropolitan peripheries are also places where the contestations and fragmentation of information on land and layer are most pronounced, and municipal leverage over information flows is clearly heightened.

Informational infrastructures of the platform

Scholarship across digital and urban geographies in the last decade or so has increasingly focused on the ways in which governance across all scales is now mediated by platforms, and the impact this has on our everyday lives (Keith and de Souza Santos 2021; Anttiroiko 2016; Verhoeff and Wilmott 2016; Constantinides et al. 2018; van der Graaf and Ballon 2019; Janowski et al. 2018; Barns 2018b; Rodgers and Moore 2018; Barns 2018a; Stehlin 2018; Krivy 2018). From financial management (FinTech) to property management (PropTech) to the gig economy, platforms have proliferated across all aspects of urban governance and are often referred to by the umbrella term GovTech (governance technologies). Janowski et al. identify this as the rise of a 'platform paradigm' (Janowski et al. 2018), which has transformed earlier governance paradigms that were driven by bureaucratic, consumeristic or participatory approaches to one that now builds upon these to generate more automated interactions with citizens.

This paradigm produces a 'government as platform' (Barns 2018a) approach in cities, which prompted Sarah Barns to call this a form of 'platform urbanism' (Barns 2018b). With the rise of smart cities and command and control centres, platforms have become key to how urban management and governance generate knowledge of and act upon urban challenges (Kitchin 2015; Kitchin et al. 2015; Kitchin et al. 2016). Inherently, platform urbanism has been linked to the encroachment of capital interests into urban development (Sadowski 2020; Leszczynski 2019; Rodgers and Moore 2018; Barns 2018a; Stehlin 2018), which can be seen in the proliferation of land management platforms (as in the case of VU) that seek to automate citizens' interactions with planning and licensing processes – that is, processes that open up land for real estate and generate revenue for the state, which some scholars have noted as the 'Digital Growth Machine' (Rosen and Alvarez León 2022).

Platforms such as VU, created for the specific purposes of managing and automating planning processes, reproduce the digital growth machine, whereby digitalisation and urbanisation go hand in hand. As digital technologies become more sophisticated, they are used to map and record metropolitan expansion, and as urbanisation spreads beyond metropolitan boundaries, digital technologies are expanded to reach beyond metropolitan peripheries. Platforms that seek to automate planning and land management are part of a broader initiative through which land in metropolitan peripheries not only enters the real-estate market, but the digitisation of this land itself becomes a political-technological terrain of contestation. As Baud et al. (2014) note, digitalisation seeks to manage spatial information to identify territories suitable for expansion. Further, urbanisation in a digital age relies on platforms such as VU to identify new territories for revenue generation (Datta 2023). As infrastructures are becoming more platformised and thus opening what Constantinides calls 'architecture and governance control points' for future urbanisation, platform infrastructures are also expanding their 'reach and scope' into the management and governance of territory (Constantinides et al. 2018). Thus, these platforms can no longer be defined only through their relationship with the city; they also aim to expand governance over those territories not earlier imagined as the 'city'. It is often in the geographic peripheries that informational expansion and urbanisation exist simultaneously – absent from urban territorial databases and yet facing rapid agrarian land transformations to join the urban land market. These territories exist both inside and outside the urban core. Inside, because new territories are key to the future of urbanisation, and outside, because they often remain beyond the reach of traditional urban governance.

Here cadastral maps remain a key tool 'in the service of the state' (Kain and Baigent 1992) in identifying new territories for capital generation and urbanisation. As Scott argues, 'the value of cadastre maps to the state lies in its abstraction and universality' (Scott 1999, 45). Krupa further notes that in the last few decades cadastral maps have become central to the neoliberal economies of the Global South, enabling states to reduce complex land relationships into simple asset ownership, making both territory and populations visible for capital (Krupa 2015). However cadastral maps are often locally created using local protocols, methods and expertise, and are therefore difficult to standardise and centralise for digital platforms (Bogaerts and Zevenbergen 2001). Further, cadastral maps are also marked by information silences, absences and gaps as they seek to draw boundaries around formally recognised properties only.

Campbell (2015) shows this in the case of Brazil's land reform programme, which presented 'vernacular dispositions of property' and ultimately became a tool in the service of elite accumulation of power. Harvey (2013) therefore concludes that while cadastral mapping is used extensively in GIS and cartography, it has varying levels of accuracy and is therefore limited in its use as a 'calculative power' of state-centric constructions of territory. In much of Mexico and indeed the world, cadastral maps are taken as the final arbiters of land ownership and therefore the exertion of state power is central to control over cadastral maps.

Any digital land information platform, such as VU, which is built upon a cadastral map base will therefore inherently produce informational peripheries by design. Following Datta (2023), we argue that informational peripheries are 'spaces of informational exclusion and fragmentation in the margins of political-informational rights to territory'. A digital platform produces informational peripheries by design when it builds upon pre-existing analogue maps with specific syntactic and informational paradigms. For example, cadastral maps (as physical records of continuously changing and mutating property relationships) are crucial to the design of digital land management platforms. However, due to their historical nature, they only exist in paper formats. The process of transferring paper-based information to digital platforms requires specific allowances and manipulations of territory across two incompatible information infrastructures. Political leverage and ownership of cadastral information also have strong connections to the ways in which information flows from the asset owners/stakeholders of this information to the platform itself. A platform needs to transfer data on property ownership and territorial boundaries to digital formats and use this information to automate future urban planning decisions. However, cadastral maps – as 'tools of statecraft' (Kain and Baigent 1992) – often exclude information relating to ambiguous land categories. This is evident in Richter's work (Richter 2011; Richter and Georgiadou 2016) in India, where she notes that digital property databases are plagued by fragmented information flows and absent data. The challenges of land information platforms such as VU are therefore twofold. First, they are reliant on both external and internal information flows within and beyond the state. Second, these flows are shaped by temporalities of socio-political structures of governance vested in different state scales and institutions, along with a diversity of juridical-technological instruments for state governance.

As Constantinides explains, '[b]eyond technology-related issues, a platform's governance also includes incentive structures that influence the strategic behaviour of stakeholders and the monetary profits of the platform' (Constantinides et al. 2018, 384). He further notes that 'control mechanisms let platform owners enforce rules that reward and punish behavior, and establish best practices on the platform' (Constantinides et al. 2018, 384). Here Bratton's work on the 'stack' is significant for explaining how platforms such as VU produce informational peripheries by design. Bratton notes that platforms are created through the layering of information from divergent sources over one another to assemble what he calls the 'stack', or an accidental superstructure of information. For Bratton, closures and borders are interdependent – a border somewhere will close off access elsewhere. This production of a stack means that most urban platforms, rather than being 'heterogeneous and open interfacial platforms, prioritize instead urban scale walled gardens' (Bratton 2015a, 36). Walled gardens both redline and exclude specific informational spaces by design. VU can thus be seen as a superstructure stack that reinforces existing unevenness in digital infrastructures and capacities across a metropolitan region, creating a bounded space with closed limits.

VU, therefore, is reliant upon the internal rules and architecture of layering information into a digital superstructure. These infrastructural aspects of platform architecture reveal VU's land information system as a continuously changing territoriality (Kitchin 2023). It is 'grafted' over existing infrastructures of information (Mattern 2017) and layered across several scales and spaces of information to create a 'stack' (Bratton 2015b). This is particularly salient when it comes to the three axes of land, layer and leverage, since jurisdictional reach over land and its boundaries changes over time, cadastres mutate as property changes hands or is inherited, and changing laws, policies and regulation determine the type of leverage that can be exerted by the state and state officials at different scales. Informational peripheries are, then, built into VU by design as land, layer and leverage intersect and coalesce over time, across geographies and scales, making and remaking the digital terrain of metropolitan urbanisation.

Methodology

This chapter developed from the five-year 'Regional Futures' project, which examines the transformation of the metropolitan peripheries of Nairobi (Kenya), Guadalajara (Mexico) and Mumbai (India) through the digitalisation of territorial governance. The discussion in this chapter is based on nine semi-structured interviews conducted since November 2022 with Jalisco government officials and municipal officers, civil society actors, land professionals and land brokers in the GMA. Key sources for this chapter are semi-structured interviews conducted with officials engaged in the design of VU at the geomatics offices of national innovation at the GMA, as well as officials engaged in cartography in the cadastre offices at the municipal level; officials of parastatal institutions (such as the Institute of Statistical and Geographical Information of Jalisco (IIEG) and the Institute for Development Planning and Management of the Guadalajara Metropolitan Area (IMEPLAN)) at the sub-national level; and officials of the National Institute of Statistics and Geography (INEGI) at the national level. They were either directly involved in some aspect of land digitisation programmes or had first-hand knowledge and experience of the production of VU.

These interviews and engagements with multiple agencies were facilitated by the strong professional and research networks of the first author, which included connections with national, sub-national and municipal government officials and land professionals. We also used snowballing techniques to identify and approach additional actors. Our interviews sought to explore the processes of digitisation within the different land administration departments and how they were assembled into automated platforms in the GMA. Most of the officials were open to interviews and keen to contribute to discussion of the VU platform, expressing pride at being able to execute the complicated task of assembling an automated platform for the region. All audio interviews were recorded, transcribed and coded. On the rare occasions officials did not consent to recording, we made copious field notes, which also became part of the research data. Most interviews were conducted in Spanish, although a few officials also spoke some English to engage with the second author. Transcripts were translated and regularly reviewed by the research team to ensure consistency of language and cultural expressions.

Land digitalisation as a 'federal pact'

Mexico is a federal republic comprising 31 states and Mexico City. The federal pact defined in the 1917 constitution considered only two scales of sovereignty: the federation and the states. It also defined the municipality as the basic territorial, political and administrative unit. From 1921, a process of centralisation began that reached its final stage in around 1982, when federal institutions were found to be inadequate in ensuring the social welfare of the population (Rodríguez 1998). This forced a change of course towards a new federalism, accompanied by

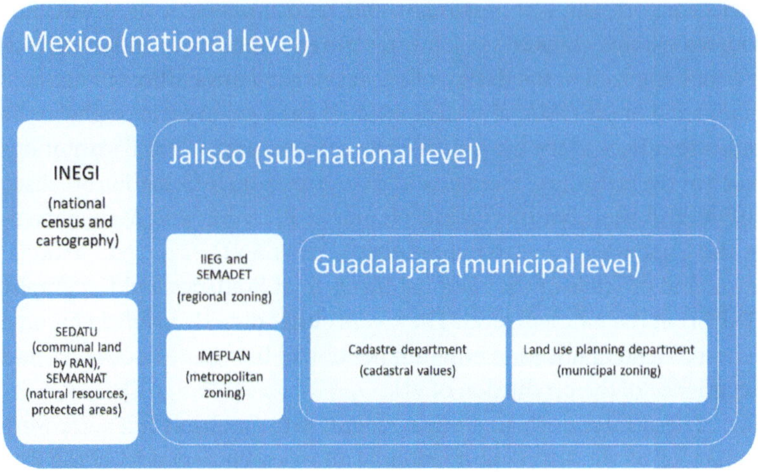

Figure 4.1 Institutions that produce cartography for various levels of government in Mexico. Source: created by the authors.

important reforms in Mexico, including the devolution of powers from the federal government to local governments in 1993. As a result, public policies are now defined at the federal level, and municipalities are delegated responsibility for implementing those policies (Bernstein 1993; Ward and Rodríguez 1999). In 1976, municipalities became responsible for land planning and management; nevertheless, these duties were not supported by funding for building capacity or facilitating the process in local agencies.

Figure 4.1 illustrates the principal institutions responsible for producing territorial information at the three levels of government in Mexico. At the federal level, the National Institute of Statistics, Geography and Informatics (INEGI – Instituto Nacional de Estadística de Geografía e Informática, created in 1983), the largest producer and custodian of data at the national level, produces census data, economic activity data and most of the layers available for cartography. Additionally, other federal ministries, such as the Secretariat of Agrarian, Land and Urban Development (SEDATU – Secretaria de Desarrollo Agrario, Territorial y Urbano) and the National Agrarian Registry (RAN – Registro Agrario Nacional), focus on urban development data and common land tenure. Also at the federal level, the Ministry of Environment and Natural Resources (SEMARNAT – Secretaria de Medio Ambiente y Recursos Naturales) surveys the physical environment and biodiversity.

At the sub-national level, the Institute of Informatics, Statistics and Geography (IIEG – Instituto de Informática, Estadística y Geografía) is responsible for the administration and generation of geographic, statistical and population information, with the support of secretaries at this level of government. The State of Jalisco Secretariat of the Environment and Territorial Development (SEMADET – Secretaria de Medio Ambiente y Desarrollo Territorial) is responsible for land planning and management for the state of Jalisco; therefore, it is focused on updating data on municipalities, although it usually does not include GMA in its scope. The Metropolitan Planning Institute (IMEPLAN – Instituto Metropolitano de Planeación) is tasked with the management and production of information with the intention of coordinating the nine metropolitan municipalities of the GMA. Additionally, the municipal government produces its own information through its directorates, with the Directorate of Cadastre and Land Use Planning being the most relevant for digitalising the territory. In this context, many institutions produce and manage information that sometimes overlaps, duplicates or replaces gaps at other scales (see Figure 4.1).

Currently, most government institutions have a digital platform holding territorial information. The digitisation process has been asynchronous, necessitating the implementation of distinct projects at various levels of government. INEGI, for example, developed vectorised digital cartography based on work previously prepared by the Commission for Studies of the National Territory and Planning, which mapped Mexican territory from its inception in 1968 until 1980, when it became the Directorate of Geography and later INEGI. This information has been adopted by sub-national and municipal governments throughout the country.

In addition to the censuses conducted primarily in urban areas, INEGI has participated in other projects to cover the remainder of the territory. Among these projects is the Programme for Certification of Rights to Ejido Lands (PROCEDE), which ran from 1993 to 2006. This initiative provided a foundation for mapping communal land (see below).

At the sub-national level of government, the state of Jalisco has a long history of aerial photography surveys dating back to the 1950s. As part of the decentralisation process, digitisation took place between 1996 and 1998, when the state government implemented the 'Digitising Jalisco' project at the Institute of Statistical and Territorial Information of Jalisco, which in 2013 became the IIEG, a parastatal organisation similar to INEGI at federal level, formed by the merger of three previously independent government bodies. That project aimed to transfer the 1:500,000-scale cartography from paper to vector format. Photogrammetric flights were also carried out to update the cadastral maps of the larger municipalities of the GMA in Guadalajara, Zapopan, Tlaquepaque and Tonala, as well as 74 other medium-sized cities. However, despite the provision of equipment by the state government, not all municipalities and cities had the capacity to use mapping to integrate and link all their cadastral information.

Other digital platforms in Jalisco, such as Mapa Jalisco y SIGmetro, started by digitising earlier paper-based maps and AutoCAD files with land information but gradually migrated these to GIS. From 2014, IMEPLAN collected municipal cartography and data, mostly from municipal urban plans. However, IMEPLAN officials noticed significant disparities in the information generated by each of the municipalities. Few municipalities had accurate information about their territory, and others had incomplete and inaccurate data. Faced with this challenge, IMEPLAN took the initiative in 2015 to develop a digital platform called SIGmetro. This platform aimed to homogenise the information of the nine metropolitan municipalities.

Since 1993, district and subdistrict plans for the municipality of Guadalajara have been developed using photogrammetry from aerial surveys commissioned by the state government. The aerial photography and its digital restitution enabled the conversion of paper plans into digital versions using computer-aided design (CAD) software. Although the municipality of Guadalajara has had a digital cadastre since the 1990s, only in the 2000s was the cadastre migrated to a GIS format (GU240607I070). In addition, the municipality has a platform called Mapa Guadalajara, managed by the Geomatics Office. This platform integrates all the information produced by the municipality's generating areas, such as infrastructure, trade, services, risks, population, and satellite and LIDAR imagery.

Besides Guadalajara, other municipalities also made efforts to digitalise territory. Zapopan was the first municipality of the GMA to migrate its urban development plans in 2010. As a result, the plans published in 2012 were the first of their kind to be created entirely from scratch in a GIS platform. However, Zapopan also inherited the legacy of informational gaps that remain when paper maps are scanned to CAD systems and later overlaid with GIS information. Furthermore, the GIS plans did not integrate cadastre information (as it was still on paper), which was essential if an integrated platform of geo-referenced data for the municipalities was to be achieved.

Visor Urbano: informational peripheries by design

VU was seen as a tool to combat corruption, enable evidence-based decision making and promote cost-efficiencies for the government and its citizens (Arauz et. al 2017; Rivero del Paso 2020; Mellon and Lara-García 2020). As the first digital platform of its kind in the municipality of Guadalajara, VU attempted to curate territorial and property data, not only to boost tax revenues, but also to manage future urbanisation. The platform's key transformations were claimed to be in the areas of regulatory overhaul, standardisation and automation, which would reduce human error or 'discretion' (in other words, corruption). VU was also citizen-facing, in that it would provide a portfolio of online information on any property (including information on zoning and primary physical characteristics) to facilitate construction licence applications.

Initially, the municipality's Land Use Planning Office was responsible for migrating the urban plans to GIS. Later, the Geomatics Office, part of the Directorate of Government Innovation, created a digital platform to publish and automate them. This involved a year of preparatory work, moving data from one information infrastructure to another – migrating plans from a CAD platform to GIS. However, information flowed in one direction: from the offices that generated the plans to the Geomatics Office which assembled them, with few opportunities for verifying or updating the information transferred to the platform. In effect, when the VU platform was launched in 2018, it was built using information inherited from decades of earlier CAD digitisation. This presented challenges to the Geomatics Office as the data were coming from several sources and were not necessarily harmonised with each other, yet they had to find a way to make all the data interoperable to build VU. As one official explained, despite being a newly created innovation agency, they were kept 'at the top in this coordination above the secretariats and strategic cabinets and belonged to an area called the Governor's Cabinet'. And therefore, despite being a Guadalajara municipality initiative, VU was scaled up to become part of Jalisco's Digital Strategy.

> [T]he digitising process was conducted by a group of young people. And we built the base on which they, you know ... digitised all the zoning and what we do is use this first information we have translated from AutoCAD and PDF to GIS formats, we combine it with the information we have on the cadastre to match exactly. And that was the base where everything was built. They had set up a central database where all the things were working at the same

> time ... [S]ome people said, this can't be done, you don't have [the] experience. (GS202)

Through the digitalisation of urban development plans and licences, VU enabled online consultations on land use, business operating licences, the renewal and reissue of small construction licences, and reporting for construction irregularities (Arauz et. al 2017; Rivero del Paso 2020). By 2019, eleven thousand consultations on land use had been carried out on VU. This was seen as a success by Bloomberg, and the Geomatics Office was asked to scale up the platform to cover a wider territory and a diversity of licensing processes.

> From 2016 to 2020, Visor Urbano worked perfectly ... And Bloomberg said: 'Hey, it was incredible everything you've had in construction licences, what if we do more things and not just in the city.' So, in 2020, that's when we started Visor Urbano, but throughout the state ... Continuing with open government issues, transparency, citizen participation ... So, we said, um, urban development is important, it also has to do with business licences. They go hand in hand a lot. And that's why we started to digitise the business licence process as well as the construction process. (GI014)

So, from 2020, the VU project entered a new phase, moving to an innovation directorate of the Jalisco state government similar to that of the municipality of Guadalajara. A dedicated team was formed with the goal of promoting the platform and replicating it in all the municipalities of Jalisco and even in other states of Mexico. Today, VU is used in 33 municipalities in Jalisco, two municipalities in the state of Nuevo León, and at state level in Hidalgo and Morelos states.

VU Guadalajara was the first version of the programme and is the most fully developed, with its entire process of licensing now semi-automated thanks to the vast amount of information that has been mounted on the platform. VU also has two offices: one within the Public Works Directorate, responsible for building permits, and another in the Directorate of Registration and Licences, handling business procedures. Figure 4.2 shows the basic structure of VU Guadalajara. Several layers of information sources feed it:

- economic units (business types) and their location, presented in the National Statistical Directory of Economic Units produced by INEGI;

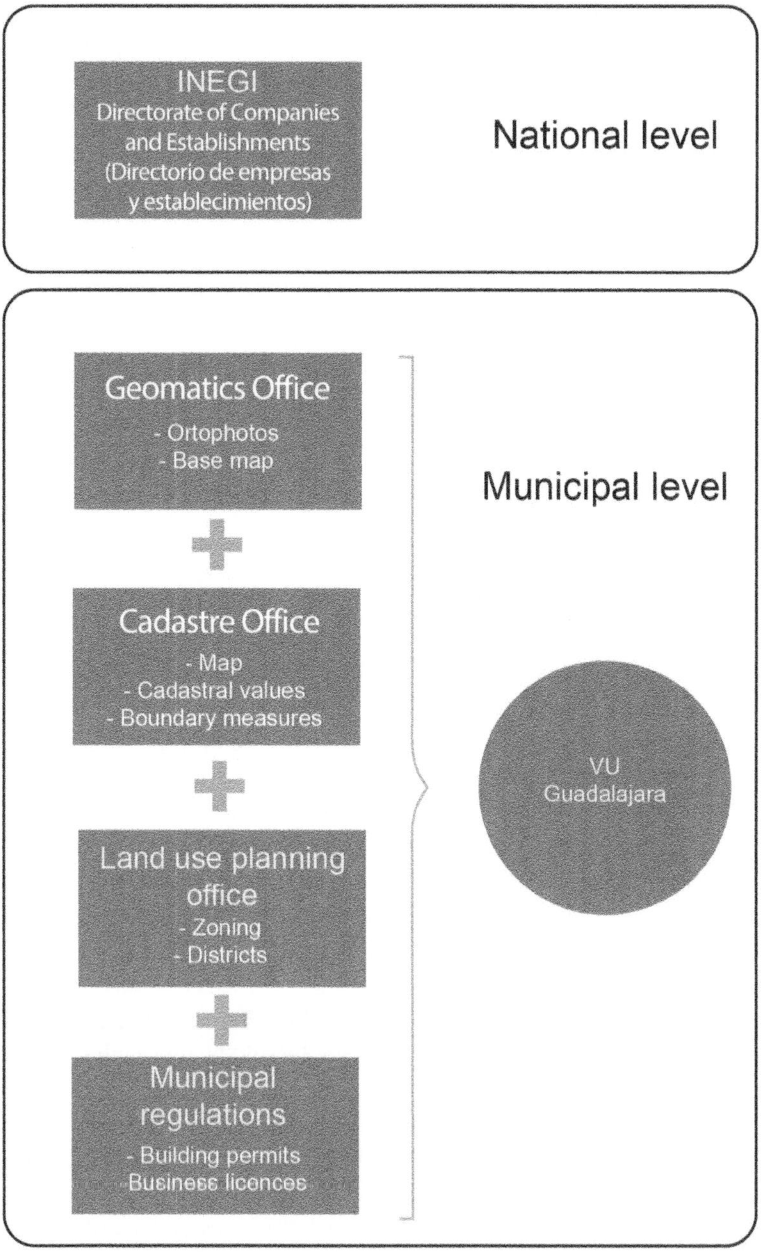

Figure 4.2 The structure of Visor Urbano Guadalajara. Source: created by the author.

- orthophotos purchased by the municipality, providing a photographic representation of the terrain and a base map with detailed and updated information on municipal offices (urban trace, blocks, road network, etc.) based on INEGI's geo-statistical framework but elaborated by Guadalajara's Geomatics Office;
- cadastral maps showing property boundaries and building floors, cadastral values and boundaries developed by the municipal Cadastre Office;
- zoning plans and polygons of municipal districts proposed by Guadalajara's Land Use Planning Office.

All these layers interact with each other in each step of the process. The platform analyses a citizen's application and issues a pre-approval if it is compatible. Then the official procedure begins, and all the documents must be collected. The process ends with a digital file uploaded and checked by the municipality's staff.

Land: ambiguous legalities

While the main objective of VU was to digitise land for automating construction licences, it could only digitise land that was formally and legally owned. This is where the geographic urban peripheries of Guadalajara overlapped with the informational periphery of VU, since a sizeable portion of land located on the peripheries of the GMA was customary or commons land that belonged to two categories of social property – *ejido* land, and land belonging to *comunidades indígenas* (indigenous communities) – which are not formally incorporated within municipal jurisdictions. *Ejido* land, created after the Mexican Revolution in 1910 to rectify historic land inequalities, consists of commons land (legally owned by the national government), which is registered under the National Agrarian Registry and cannot be sold by those inhabiting it (Perramond 2008). Indigenous communities' land is formally recognised and collectively owned indigenous land, with records often going back to the colonial Spanish Crown. Since the Mexican constitutional reforms of 1991–2, *ejido* land can be bought or sold on the open market and can therefore be 'privatised' in the land market (Jones and Ward 1998). Further, *comunidades* land can be registered as *ejido* land, enabling it to enter the land market (Smith et al. 2009). As Varley and Salazar (2021) argue, that has deepened the land crisis in Mexico in recent years and has led to increased and rapid peri-urbanisation and neoliberal corporatisation.

Due to their complex legal and jurisdictional frameworks, the registration of land in *ejido* territories has been the cause of numerous tensions and conflicts. *Ejido* and indigenous communities landowners operate an independent land administration system and issue documents to their inhabitants. However, the legality of the *constancia de posesión* or *constancia de derechos* (proof of property) documents issued by the community offices for people living on communal land can always be challenged in the agrarian courts.

Krupa notes that the insistence on indigenous people obtaining property rights by moving into the formal market is 'to mark people as living with the stain of living at the margins of bureaucratic governance, as located somewhere beyond the rule of law' (Krupa 2015, 103). This is reflected in the unfinished work of PROCEDE, an extensive programme aimed at mapping social property in urban areas. PROCEDE's fundamental objective was to provide *ejidos* with certificates of usufruct rights, rights of proportional usufruct over communal lands, and titles of ownership for plots that were already within an urban area. This programme also catalogued *ejido* properties in the cities, particularly those in the peripheries of the metropolises. However, PROCEDE also left open many areas, such as land for urban growth, common use or assembly areas, as *areas sin delimitar el interior* (areas without mapping the interior), and hence without a designated land use category.

Land here is juridico-technological, in the sense that legal instruments which seek to liberalise land for the market also impose antagonistic (in/formal, il/legal) land categories. As both *ejido* and *comunidades* lands are historically produced, collectively owned and rooted in rural initiatives, they lie beyond the jurisdiction of urban municipalities, which do not hold any information on them. Thus, even though PROCEDE mapped commons land on a cadastre, these data were not shared with the urban municipalities and, therefore, did not feed into the VU platform. Land, then, as a political-technology of power (Elden 2010), seeps into the ontologies of digital information systems; its legacy as a legislative tool in the service of capital produces its own faultlines of inclusion and exclusion in and through the platform.

There is a paradox here. While the VU platform covers the areas where *ejido* and *comunidades* lands are located, the cadastre is not precise in these areas. Although the municipality holds this information more accurately through aerial photogrammetry, it is the specific protocol of data entry into the VU that disrupts informational flows from the appropriate institutions to the platform. This means that any transactions on *ejido* or *comunidades* land remain obscured from view in the platform.

Layer: a cadastral life connection

As one official put it, creating VU was like 'solving a puzzle' with pieces of fragmented and conflicting information from different state custodians at different scales. The key to this territorial information 'puzzle' can be honed down to a single data point on which the entire platform is built – the cadastral GIS layer. The cadastral layer is digitised from the earlier cadastre maps that were created on paper or CAD. The officials noted that for VU, the cadastre 'is the only dynamic layer for everything' (GS202) – in other words, it is the foundation of the VU platform. As a conduit of informational flows, the cadastral layer needs to be continually updated on the platform in real time to maintain its usefulness. The cadastre map, in the words of another official, is 'a life connection' for metropolitan cartography.

However, despite being central to the viability of the VU platform, not all of the nine municipalities in the GMA had created a cadastre; those that had one had not digitised it in the correct format for uploading to the VU platform. This created uneven information flows, which were reflected in the gaps in territorial information on VU. Municipalities that were well funded, had grown recently or could generate higher revenues were better equipped to map and digitise their cadastral data, while others were left out of the VU database. As one official noted regarding one of the most recently developed municipalities in the GMA:

> Tlajomulco [municipality] has been, it was for a long time, the most advanced municipality in that regard. So, they had everything that was mapped. Even little things, also mapping of the street markets. They have a map of where each shop was located. Yeah, they got to that level of [detail] ... Because they were ... collecting taxes on the area, of the market. That's their way of seeing stuff. Map everything so we can collect taxes on everything. (GS202)

As the municipalities differed in their resource capacities, so too did the level of detail in the digitised cadastre maps shared with the Geomatics Office. In some cases, municipalities had either failed to collect the necessary data or had not represented it in their cadastre maps. This presented a challenge for the VU platform, which the geomatics officials had to find creative ways to address. They called this 'the power of the mouse'.

> Many municipalities, out of the 125 we have, don't have their cadastre. So, when that information that is useful to us to start a process or something doesn't exist, we create tools like the drawing tool, or we pull other resources like Google's [map] ... And actually, the Visor Urbano team has done this for all the municipalities. They have matched the limits of the cadastre with the zoning, which is not the best practice because these are two different things. But most of the time nobody notices, there's no issue. You can't actually just move the borders of the zoning polygon. It's not a thing. But it's the only thing we can do with this. (GS202)

This practice of compensating for a lack of cadastral information became a recurring approach throughout the development and maintenance of VU. As officials noted, this is not considered 'best practice', as it can lead to inconsistencies between cadastral and zoning data. Yet, in the absence of detailed maps from municipalities, it was the only viable solution. Here the computer mouse performs what is known in GIS as a 'fudge' or a digital compromise, where small mismatches between the lines of property and zoning are accepted as part of the platform's error tolerance. The cadastre here is a political-technological domain within the GMA, where its transformation into a digital platform enhances its networked capabilities, influencing economic and social aspects of land-based relations.

Leverage: municipal autonomy and custodianship

Even as the cadastral map emerges as the core foundation upon which any digital land information platform can be built in Mexico, it remains under the custodianship of municipalities, which can choose what cadastral information to share. This is due to ambiguous changes in concurrent laws such as the General Law of Human Settlements, Territorial Management and Urban Development, leaving municipalities with relative autonomy as to what and how much data they share with other state institutions at different scales. Although municipalities have innovation coordination offices and geomatics headquarters responsible for collecting and updating information from various municipal offices, they still do not always allow the exchange and feedback of data vertically with state and federal institutions. This issue is particularly pronounced in peripheral municipalities like Zapopan, which has long maintained strict data privacy measures. Although Zapopan was the first municipality to develop a GIS-based urban development plan in 2012, it did not integrate cadastral information within its plans. Zapopan has always kept cadastre information protected and has not integrated it with other aspects of territorial digitisation.

> Zapopan is difficult. They have an NDA [non-disclosure agreement]. They ask you to sign an NDA if you want to watch or look at the cadastral data. They work at that level. You know … impede access to the data. Well, they had a lot of issues where they had to do the cadastral cartography … They had done several versions of the cartography, and all the versions had been a failure. So, they keep it closed so nobody can see the mess they have inside. (GS202)

Another official referred to the situation as follows:

> Until now we haven't provided anything. Look, right now I think the reason is mostly because of the level of privacy. I don't know how much information … should [be] release[d]. So, in the topic of privacy, we are very careful. (GI012)

The challenges around sharing data vertically between geomatics offices and municipalities further complicated the prospect of maintaining VU and keeping it relevant. If, under the federal pact, municipalities had autonomy over data sharing, then the foundation of the VU platform built on cadastre maps was designed to fail. Further, because of this autonomy, municipalities could implement their own platforms and digital land governance systems in parallel to VU. As each municipality also had a different regulatory framework, this meant that the licensing system supported by VU had to be customised for use by each municipality. As one official noted:

> [F]rom our side we have to respect the independence of the municipalities as well. There are some municipalities that already have their software. They are already using it for licences or cadastre, right? So, there is no longer a need. There are others that are clearly paying for them. There are others who are afraid of this important challenge of digitalisation, or see it as a big … it makes them a bit nervous, right? (GU230427I014)

The larger objective of VU, officials noted, was to bring municipalities into the digital era through soft power incentives. The Geomatics Office thus launched a project called 'Municipalities of the Future', where officials would be trained in accessing technologies and ethically responsible uses of data. They called this 'ecosystem development' by bringing in private companies or tech groups to share knowledge and expertise.

> [T]he good thing about this Visor Urbano is that it's support that the state government and Bloomberg offer to municipalities. So, the municipalities don't pay anything at all, that's why. Their information is only theirs; we don't have access. They manage it, they update it, they configure it, and it's loaded on the state's servers here. It's like a co-investment between Bloomberg and the Jalisco state government so that Visor Urbano can be used. That's why my premise is that Visor Urbano really works more for slightly smaller municipalities. (GI014)

This observation about smaller municipalities being better suited to VU is related to the digitisation of maps, the volume of data under municipal custodianship, the capacities of municipalities and their relative autonomy to share data. Smaller municipalities are usually more reliant on support from the Jalisco state government and its Geomatics Innovation Office and, therefore, are more conducive to sharing data on VU. Thus, for the viability of a digital platform such as VU, size matters inversely.

The larger the municipality, the more independent and less reliant it will be on data sharing, increasing the likelihood of informational gaps in the platform. The concept of metropolitan coordination is challenging to implement due to the autonomy and political leverage of the most populous municipalities in the GMA. Since their budgets allow them to be independent, it is hard for all the municipalities to agree on a single set of criteria to build a metropolitan platform.

> [T]his has been one of the main challenges since the beginning of the [VU] project. When we applied for the Mayor's Challenge in 2016, one of the conditions for the award was to have an idea that could be replicated in other cities. And our proposition was to replicate VU in the metropolitan area. So, our first goal in that year was to have a metropolitan Visor Urbano, but the reality was really complicated. The political issues have a second life, or they need to talk a little bit further about that. (GI054)

Here it is worth referring to Saskia Sassen's suggestion that territory needs to have a conceptual autonomy from territoriality. Sassen notes that ‘territory, here understood as a complex capability with embedded logics of power/empowerment and of claim making, some worthy and some more akin to power-grabs’, is something that is different from how the nation state, or the state at any level, asserts its sovereignty (Sassen 2013, 21). In the case of VU, however, territory exists on at least two levels: the conventional material political-legal landscape of power, and the virtual or representational power of this landscape on the platform. Territoriality here is exercised by the municipal state over both material and virtual landscapes, through its leverage over sharing cadastral information with the platform. The larger a municipality, the stronger its claims-making ability regarding the virtual representation of its territory on the platform. This sort of claims making realigns the material and virtual territory of the GMA to produce new types of informational peripheries that are tied to the protocols of land platformisation.

VU and informational peripheries

In this chapter we examined how informational peripheries are designed into the structure of automated land information systems. We suggested that the three political-technological spaces of land (as a physical asset), cadastral layer (as a conduit of information flows) and leverage (as autonomy in data sharing and access) produce informational peripheries within the digital platform. These informational peripheries are grafted onto the geographic peripheries of the metropolitan region, which are already marked by juridico-technical ambiguities in land, data absence or scarcity, and multiple/fragmented custodians of information.

The aspirations of VU as an automated, centralised digital platform for future land governance are in tension with the future of de/centralised power of the state whose ‘magic’ (Taussig 1997) emanates from democratising different layers of state governance. Decentralisation has been the bedrock of democratic innovation for several states in the Global South, yet, as Krupa (2015) suggests, this recalibrates the state’s relationship with land as a political-technological relationship and the cadastre as a tool of statecraft. Decentralisation influences how land is to be counted, by which level of state, and how that information will flow vertically. A digital platform by design requires centralised custodianship of data as the key to seamless automation within it, and if information does not flow in centralised, standardised formats, it means that any platform will always produce informational peripheries by design.

To understand informational peripheries, as the introduction to this volume suggests, local governments that have embarked on developing digital platforms such as VU need to rethink the geographical notion of periphery through the capacities afforded by digital infrastructures and technologies. We argue here that the informational periphery is both political-technological and historically produced, shaped by both inertia and the flows of data across geographical boundaries of the urban and across state spaces. Informational peripheries are grafted onto geographic peripheries through varying levels of data gaps/absences/ scarcities that exist in the far-flung metropolitan regions that have historically been left out of 'urban' cartography, even though these are now the frontiers of metropolitan urbanisation. Here 'inertia' refers to the resistance or reluctance of state institutions to share their data, producing information gaps in the platform. 'Flows' refers to the way data is shared, which means it may not be up to date or in the correct format to be useful for automation in the platform. In the GMA, inertia and flows of data are inherently geographic as they are bottlenecked across municipal and scalar boundaries of the state. As Agrawal and Kumar (2020) argue, 'who counts, what counts and how to count' remain inherently political choices in state institutions responsible for cartography that inherently turn the platform into a political-technological reflection of statecraft.

Acknowledgements

Funding for this research was provided by the European Research Council Advanced Grant 'Regional Futures: The territorial politics of digitalisation-as-urbanisation in the Global South' (ID: 101019318) awarded to Ayona Datta. We are grateful to Fenna Hoefsloot, who conducted some of the interviews alongside the authors. We also want to thank all the professionals and organisations who participated in this research and shared their knowledge and time with us.

References

Agrawal, Ankush and Vikas Kumar. 2020. *Numbers in India's Periphery: The political economy of government statistics*. Cambridge: Cambridge University Press.

Anttiroiko, Ari-Veikko. 2016. City-as-a-platform: The rise of participatory innovation platforms in Finnish cities. *Sustainability* 8(9): 922. https://doi.org/10.3390/su8090922.

Arauz, Mario R., Yunive Moreno, Raúl Nanclares, C. Vanessa Pérez and Victor M. Larios. 2017. Tackling corruption in urban development through open data and citizen empowerment: The case of 'Visor Urbano' in Guadalajara. International Smart Cities Conference (ISC2). https://doi.org/10.1109/ISC2.2017.8090840.

Ash, James, Rob Kitchin and Agnieszka Leszczynski (eds). 2018. *Digital Geographies*. Thousand Oaks, CA: Sage Publications.
Barns, Sarah. 2018a. Smart cities and urban data platforms: Designing interfaces for smart governance. *City, Culture and Society* 12: 5–12. https://doi.org/10.1016/j.ccs.2017.09.006.
Barns, Sarah. 2018b. Platform urbanism rejoinder: Why now? What now? *Mediapolis: A journal of cities and culture*, 7 November. Accessed February 2025. https://www.mediapolisjournal.com/2018/11/platform-urbanism-why-now-what-now.
Baud, Isa, Dianne Scott, Karin Pfeffer, John Sydenstricker-Neto and Eric Denis. 2014. Digital and spatial knowledge management in urban governance: Emerging issues in India, Brazil, South Africa, and Peru. *Habitat International* 44: 501–9. https://doi.org/10.1016/j.habitatint.2014.09.009.
Bernstein, Tao Neil. 1993. Fifty years of state governors in México: Middle elites and political stability. PhD diss., University of Virginia: Woodrow Wilson Department of Government and Foreign Affairs.
Bogaerts, T. and J. Zevenbergen. 2001. Cadastral systems: Alternatives. *Computers, Environment and Urban Systems* 25(4): 325–37. https://doi.org/10.1016/S0198-9715(00)00051-X.
Bratton, Benjamin H. 2015a. Cloud megastructures and platform utopias. In *Entr'acte: Performing publics, pervasive media, and architecture*, edited by Jordan Geiger, 35–51. New York: Palgrave Macmillan.
Bratton, Benjamin H. 2015b. *The Stack: On software and sovereignty*. Cambridge, MA: MIT Press.
Campbell, Jeremy M. 2015. The land question in Amazonia: Cadastral knowledge and ignorance in Brazil's tenure regularization program. *PoLAR: Political and Legal Anthropology Review* 38(1): 147–67. https://doi.org/10.1111/plar.12091.
Constantinides, Panos, Ola Henfridsson, and Geoffrey G. Parker. 2018. Introduction: Platforms and infrastructures in the digital age. *Information Systems Research* 29(2): 381–400. https://doi.org/10.1287/isre.2018.0794.
Datta, Ayona. 2023. The digitalising state: Governing digitalisation-as-urbanisation in the Global South. *Progress in Human Geography* 47(1): 141–59. https://doi.org/10.1177/03091325221141798.
Elden, Stuart. 2010. Land, terrain, territory. *Progress in Human Geography* 34(6): 799–817. https://doi.org/10.1177/0309132510362603.
Harvey, Francis. 2013. The power of mapping: Considering discrepancies of Polish cadastral mapping. *Annals of the Association of American Geographers* 103(4): 824–43. https://doi.org/10.1080/00045608.2012.720228.
Janowski, Tomasz, Elsa Estevez and Rehema Baguma. 2018. Platform governance for sustainable development: Reshaping citizen-administration relationships in the digital age. *Government Information Quarterly* 35(4): S1–16. https://doi.org/10.1016/j.giq.2018.09.002.
Jones, Gareth A. and Peter M. Ward. 1998. Privatizing the commons: Reforming the ejido and urban development in Mexico. *International Journal of Urban and Regional Research* 22(1): 76–93. https://doi.org/10.1111/1468-2427.00124.
Kain, Roger J. P. and Elizabeth Baigent. 1992. *The Cadastral Map in the Service of the State: A history of property mapping*. Chicago, IL: University of Chicago Press.
Keith, Michael and Andreza Aruska de Souza Santos. 2021. *African Cities and Collaborative Futures: Urban platforms and metropolitan logistics*. Manchester: Manchester University Press.
Kitchin, Rob. 2015. Making sense of smart cities: Addressing present shortcomings. *Cambridge Journal of Regions, Economy and Society* 8: 131–6. https://doi.org/10.1093/cjres/rsu027.
Kitchin, Rob. 2023. *Digital Timescapes: Technology, temporality and society*. Cambridge: Polity Press.
Kitchin, Rob, Tracey P. Lauriault and Gavin McArdle. 2015. Knowing and governing cities through urban indicators, city benchmarking and real-time dashboards. *Regional Studies, Regional Science* 2(1): 6–28. https://doi.org/10.1080/21681376.2014.983149.
Kitchin, Rob, Sophia Maalsen and Gavin McArdle. 2016. The praxis and politics of building urban dashboards. *Geoforum* 77: 93–101. http://dx.doi.org/10.2139/ssrn.2608988.
Krivy, Marcos. 2018. We are all platform urbanists, but not all in the same way. *Mediapolis: A journal of cities and culture* 3(4). Accessed February 2025. https://www.mediapolisjournal.com/2018/11/platform-urbanists-not-in-the-same-way.
Krupa, Christopher. 2015. Cadastral politics: Property wars and state realism in highland Ecuador. In *State Theory and Andean Politics: New approaches to the study of rule*, edited by Christopher Krupa and David Nugent, 99–125. Philadelphia, PA: University of Pennsylvania Press.

Leszczynski, Agnieszka. 2019. Glitchy vignettes of platform urbanism. *Environment and Planning D: Society and space* 38(2): 189–208. https://doi.org/10.1177/0263775819878721.
Mattern, Shannon. 2017. A city is not a computer. *Places Journal*, February. https://doi.org/10.22269/170207.
Mellon, Jonathan and Francisco Lara-García. 2020. *Evaluation of the Guadalajara Visor Urbano Commercial Permitting System's Effect on Corrupt Practices*. Accessed February 2025. https://visorurbano.guadalajara.gob.mx/formatos/Visor_Urbano_Impact_Evaluation_Delivery.pdf.
Perramond, Eric P. 2008. The rise, fall, and reconfiguration of the Mexican ejido. *Geographical Review* 98(3): 356–71. https://doi.org/10.1111/j.1931-0846.2008.tb00306.x.
Richter, Christine. 2011. In-tensions to infrastructure: Developing digital property databases in urban Karnataka, India. *Environment and Urbanization Asia* 2(2): 205–22. https://doi.org/10.1177/097542531100200205.
Richter, Christine and Yola Georgiadou. 2016. Practices of legibility making in Indian cities: Property mapping through geographic information systems and slum listing in government schemes. *Information Technology for Development* 22(1): 75–93. https://doi.org/10.1080/02681102.2014.886548.
Rivero del Paso, Lorena. 2020. *Govtech and the Future of Government: The case of Visor Urbano in Mexico*. Caracas: CAF.
Rodgers, Scott and Susan Moore. 2018. Platform urbanism: An introduction. *Mediapolis: A journal of cities and culture* 3(4). Accessed February 2025. https://www.mediapolisjournal.com/2018/10/platform-urbanism-an-introduction.
Rodríguez, Victoria E. 1998. Recasting federalism in Mexico. *Publius: The journal of federalism* 28(1): 235–54. https://doi.org/10.1093/oxfordjournals.pubjof.a029950.
Rosen, Jovanna and Luis F. Alvarez León. 2022. The digital growth machine: Urban change and the ideology of technology. *Annals of the American Association of Geographers* 112(8): 2248–65. https://doi.org/10.1080/24694452.2022.2052008.
Sadowski, Jathan. 2020. Cyberspace and cityscapes: On the emergence of platform urbanism. *Urban Geography* 41(3): 448–52. https://doi.org/10.1080/02723638.2020.1721055.
Sassen, Saskia. 2013. 'When territory deborders territoriality'. *Territory, Politics, Governance* 1(1): 21–45. https://doi.org/10.1080/21622671.2013.769895.
Scott, James C. 1999. *Seeing Like a State: How certain schemes to improve the human condition have failed*. New Haven, CT: Yale University Press.
Smith, Derek A., Peter H. Herlihy, John H. Kelly and Aida Ramos Viera. 2009. The certification and privatization of indigenous lands in Mexico. *Journal of Latin American Geography* 8(2): 175–207. https://doi.org/10.1353/lag.0.0060.
Stehlin, John. 2018. Urban platforms, rent, and the digital built environment. *Mediapolis: A journal of cities and culture* 3(4). Accessed February 2025. https://www.mediapolisjournal.com/2018/10/urban-platforms-rent-and-the-digital-built-environment.
Taussig, Michael. 1997. *The Magic of the State*. Abingdon: Routledge.
van der Graaf, Shenja and Peter Ballon. 2019. Navigating platform urbanism. *Technological Forecasting and Social Change* 142: 364–72. https://doi.org/10.1016/j.techfore.2018.07.027.
Varley, Ann and Clara Salazar. 2021. The impact of Mexico's land reform on periurban housing production: Neoliberal or neocorporatist? *International Journal of Urban and Regional Research* 45(6): 964–84. https://doi.org/10.1111/1468-2427.12999.
Verhoeff, Nanna and Clancy Wilmott. 2016. Curating the city: Urban interfaces and locative media as experimental platforms for cultural data. In *Code and the City*, edited by Rob Kitchin and Sung-Yueh Perng, 116–29. Abingdon: Routledge.
Ward, Peter and Victoria Rodríguez. 1999. New federalism, intra-governmental relations and co-governance in México. *Journal of Latin American Studies* 31: 673–710. https://doi.org/10.1017/S0022216X99005401.

5

The peripheralisation of informalised communities in Kenya's digital land administration

Fenna Imara Hoefsloot and Catherine Gateri

'Ardhi Sasa, Ardhi Tasa.' This phrase, emblazoned on banners during lawyer-led protests outside the Kenyan Supreme Court and the State Department of Lands in January 2023 (Figure 5.1), can be roughly translated as 'the land now is barren land'. It is a pun on Ardhisasa, Kenya's National Land Information Management System (NLIMS), which is currently being piloted in Nairobi.

Ardhisasa, a digital land administration platform, is a replacement for earlier paper-based land management in Kenya, which has existed since the colonial period. Ardhisasa was developed with the promise of reducing the backlog in the processing of land transactions by fast-tracking land property searches, registration, valuation and issuance of titles, culminating in accelerated investment and development of land as capital. In the longer term, this should resolve the land administration and management challenges of manual, paper-based transactions, introduce a more efficient and integrated land management system, and provide a tool to counter fraud and corruption within the land sector.

The Law Society of Kenya is just one of the many professional bodies that have expressed discontent with the newly digitalised NLIMS and its current design and functioning. Other private-sector actors, such as the Institution of Surveyors of Kenya, which represents land surveyors, land valuers, property managers and land administration managers, have directed complaints to the Ministry of Lands and Physical Planning regarding their day-to-day interaction with the Ardhisasa system and its perceived failure to properly engage with the needs and recommendations

Figure 5.1 The 'Ardhi Sasa, Ardhi Tasa' demonstration, led by the Law Society of Kenya and the Institute of Surveyors of Kenya, 20 January 2023. Source: https://x.com/lsk_nbi/status/1616450160699097089 (accessed February 2025).

of the professionals who will use and depend on this system. While the digital land information management system has generally been welcomed, both lawyers and built environment professionals argue that the implementation of Ardhisasa has led to the partial failure of services across the land registries and severe delays in land transaction processing.

The implementation of Ardhisasa represents one of the most complex and expansive reforms of Kenya's land information management system to date. It should be understood within a context characterised by bureaucratic lethargy, corruption and political patronage (Kariuki et al. 2018; Manji 2020), where the digitalisation of public services itself is a field of contestation and negotiation. In this process, as we will illustrate in our analysis, while Ardhisasa accommodates the voices of private-sector actors, it systematically marginalises the voices of civil society, particularly those advocating for peripheralised communities. However, there are many instances where peripheralised groups contest their institutionalised exclusion, find ways to manoeuvre within the structure and adapt it to meet their needs, and imagine alternative futures in the process (Kimari and Ernstson 2020). In so doing, they find new ways of writing themselves into the city, searching for loopholes in the newly enacted Community Land Act or through alternative tenure registration systems.

To unpack these dynamics, we draw attention to three distinct strategies mobilised by various actors – contestation, negotiation and experimentation – and are interested in how the three different strategies are leveraged to reconfigure how the digital platform produces informational peripheries. This chapter analyses the platformisation of urban governance (van Dijck 2021) and the peripheralisation of urban places and people (Cowan 2021; Datta 2023) to understand the production of the informational periphery in Nairobi. In doing so, it responds to calls to forge connections between these fields through the extended governance-digitalisation-urbanisation nexus (Barns 2018; Leszczynski 2020). Other empirically grounded studies have analysed digitalisation in urban contexts, either focusing on reforming governance practices (Pelizza 2017; Taylor and Richter 2017) or looking at the effects of digitalisation on the position of citizens within the city (Vanolo 2016; Calzada 2018). However, little research exists on how, particularly in southern cities, infrastructures of the digital age generate differentiated urban landscapes through the territorialisation of information systems (Hoefsloot et al. 2022; Datta 2023).

Our central thesis is that despite Ardhisasa's progressive development, it follows the tradition of a long line of large-scale infrastructural or developmental projects that rarely deliver on their promise of improvement but rather further entrench peripheralised groups (Li 2007; Anand et al. 2018; Lesutis 2022a) due to its exclusion of the already existing, albeit informalised, land administration and transaction practices that meet the needs of the urban poor. We argue that Ardhisasa's perpetual state of becoming leads to the extension of the geographical periphery into information by keeping informalised communities permanently on the threshold of becoming formalised within the digital land administration system.

Informational peripheries of platform infrastructure

Increasingly, digital platforms such as Ardhisasa are being implemented to govern state services. Previous literature on platforms and platformisation has focused on the role of big-tech companies in developing digital platforms and their extraordinary power to influence democratic governance (O'Reilly 2011; van Dijck 2021) and the organisation of public, work and social life (Artioli 2018; Heeks et al. 2021). We focus on the platformisation of government itself. Platformisation, following van Dijck (2021) and Poell et al. (2019), in this case refers to the corporate or state-controlled digital information infrastructures that have become central to collecting data, governing access to and circulation of information, and structuring interactions between users. Digital platforms are often presented as instrumental to knowing the state and guiding daily operations (Kitchin et al. 2016). Yet they regularly emerge from economically driven interests and are at odds with pre-existing practices and regulatory traditions (Poell et al. 2019) or reproduce structural inequalities because they are designed for and appropriated by the elites to serve their benefit (Senshaw and Twinomurinzi 2022). Tracing the development of Ardhisasa as a land administration platform enables us to discern its logic: how the information infrastructure developed is an enactment of the socio-political decisions made and works towards institutionalising these into space and society (Turner 1967; Bowker and Star 2000). Highlighting the process and incompleteness of infrastructure (Guma 2020), we argue that platformisation will always be an ongoing process and any notions of completeness always an aspiration. It is the internal negotiations and politics of producing these platforms that give insight into the emergence and entrenchment of informational peripheries (Datta 2023).

In this chapter we aim to capture the unsettled, ambiguous conditions that arise when the implementation of novel infrastructures creates a rupture with previous systems – this 'interstructural' (Turner 1967) moment of movement and transition, of being on the precipice of a new system. Focusing on this interstructural moment is helpful for understanding the different modes of urbanisation in the face of uncertainties and ambiguities in urban development policies, political regimes and market reforms (Müller and Trubina 2020). Müller and Trubina conceptualise this as the space where the rigid infrastructures and rationales of government meet the city's unstable, unanticipated and unruled assemblages. Highlighting the openings that are created through half-baked laws or semi-implemented plans, scholars have described how actors, from slum dwellers to elites to multilateral organisations (Lancione and Simone 2021; McConnell 2017; Müller and Trubina 2020), speculate, improvise and experiment with codes, spaces and politics as a process of 'worlding', referring to the practices through which the space itself is staged and performed (Omura et al. 2019). The never-ending accumulation of ideas, interventions and materialities keeps the city on a precipice of whatever it is amassing into (Lancione and Simone 2021).

This is particularly prevalent in the introduction and use of digital technologies for urban governance. Kitchin et al. (2016) present an understanding of urban digital technologies such as platforms, dashboards and central control rooms as 'ontogenic in nature', centring the different stages of becoming and growth these socio-technical systems undergo before reaching maturity. Whether or not a digital urban technology reaches this point of maturity and passes through the phase in between design and functioning is dependent on the alignment of state, private-sector and societal actors, technology and knowledge through the reiterative processes of visioning, negotiation, contestation and re-visioning (Kitchin et al. 2016; Kitchin et al. 2017). As we will explain, while Ardhisasa promises a radical transition in Kenya's land administration system, this has not yet been accomplished, elongating the periphery from geographical into informational space.

Agency in informational spaces

Understood this way, attuning ourselves to the different openings for reconfiguration helps us identify the strategies mobilised by actors to exercise their agency and give shape to the platform as an emerging system (Orlikowski and Scott 2021). Exploring agency within peripheral space and time can inform analysis of platform–user relationships and provide insight into how individuals and organisations mobilise to negotiate, comply with or resist the system's architecture. Within this interstructural moment, the 'rigid hierarchy' (Jordan 2015) defining how information systems are organised is mendable, and actors are less bound by their pre-existing identities and roles. Instead, there is the opportunity for individuals and organisations to carve out a new position for themselves within the socio-material system, enact new compositions between state and technology, and derive greater capabilities and outputs.

In the case of Ardhi Sasa, as existing methods of land administration are disrupted with the introduction of a digital platform, strains emerge from the suspension of routine enactments. Attending to the conditions and effects of these tensions points to both the challenges of reform and the opportunities for the emergence of alternative practice and experimentation, potentially creating opportunities for actors to intervene and reconfigure the architecture of the platform. Siles et al. operationalise this by analysing how agency in in-between spaces is present in the potential and power to act and enact change; to participate in shaping digital technologies through both resistance and compliance.

This departs from the notion that 'agency is always and simultaneously in-between spaces, positions, and worlds' (Siles et al. 2023, 64). However, this is not a level playing field. Elbanna and Idowu (2022) argue that we should be attuned to the restrictions of culture, society and politics in structuring and limiting the capacity of people to act within the system. Cultural context and regulatory frameworks impact actors' agency in shaping technological platforms for the city (Odendaal 2023).

Hence, we analyse the implementation of Ardhisasa as a space-time of negotiation, contestation and experimentation through which the informational periphery is produced. While this negotiation happens at many levels and at many times, we particularly zoom in on the influence of two actors: private-sector actors, including land surveyors, lawyers and land valuers; and civil society organisations representing informalised residents and how they exercise agency in the development of this national land information management system. Considering these two actors and their influence in the design and development of Ardhisasa, we trace the overt and covert exercise of power in the platformisation of land administration in Nairobi.

By positioning the strategies of private-sector actors in relation to those of civil society organisations, we follow Côté-Roy and Moser (2019) in their distinction between 'elite stakeholders' (such as the state and governments, multinational corporations, private foundations, non-profit organisations and global consultancy firms), who hold key roles in the shaping of policy and the mobilisation of knowledge, and subaltern actors, whose voices, knowledge and criticism are often actively peripheralised within information systems. Yet, as Li (2007) describes, critical mobilisation can come in many forms and use many strategies for social, structural and political transformation, from academics using method and writing to the many citizens or recipients of the policies who defy or challenge politics through practice.

Recognising that agency is nested within historical, political, economic and cultural contexts of colonialism, inequality and informalisation, it is important to understand these struggles to influence the development trajectory of Ardhisasa. Specifically, in the context of land administration, we should pay attention to how the digital recording of land is territorialised unevenly. As Cowan (2021) shows, this often happens through facilitating the commodification of land and the strategies of elite state and private-sector actors to capture land against the attempts of the urban poor to consolidate their property claims. Lesutis (2022b) describes this as governance through disavowal, or the way in which the state effectively pays lip service to peripheralised groups by including them in the discourse and planning of infrastructural developments while rendering them politically absent and materially unaccounted for. As a result, Lutzoni (2016) explains, informality in Kenya is not outside of planning practices but rather emerges from a relational sphere of legality, approval and negotiation, as well as contestation, eviction and delegitimisation.

Methodology

This chapter draws on 22 semi-structured interviews and observational data collected during two two-week field visits in 2022 and 2023 within the context of the 'Regional Futures' research project, analysing the process of digitalisation-as-urbanisation in the metropolitan areas of Nairobi (Kenya), Guadalajara (Mexico) and Mumbai (India). We interviewed policy makers and employees from the land administration departments of national and county governments, land surveyors, land valuers, and community organisations representing informalised residents in Nairobi.

The diversity of the researchers, being Kenyan and UK-based, and our different positionalities in the field provided interesting perspectives. In particular, we derive information from the experiences of one of the authors who has interacted with Ardhisasa in a professional capacity. Given that Ardhisasa is an ongoing project that has generated a lot of debate in the land sector, being part of the professional bodies working with the platform on a day-to-day basis gave an insider perspective into the practices and debates revolving around Ardhisasa's development. Moreover, the Kenyan researcher accepted and understood the digitisation process as a land management and administration tool from a historical perspective, having experienced the various previous land reforms in Kenya. In contrast, the UK-based researcher, with no prior experience in the Kenyan land sector, probed and interrogated the land digitisation process from a different perspective, questioning the roles of the different actors in the development of the platform. Additionally, the interviewees showed patience with the UK-based researcher – they were willing to provide detailed responses about the process – whereas the Kenyan researcher was deemed to be knowledgeable about what had been going on and thus received shorter responses.

Interviews were conducted in English and transcribed and analysed thematically, focusing on the digitisation of land records at the national government level, the different actors involved in the development of

Ardhisasa, and the challenges that arise from implementing the NLIMS in a context of extensive informality in land tenure. In addition, we attended online and in-person workshops and working sessions organised by professional bodies such as the Institution of Surveyors of Kenya and civil society organisations in which Kenya's land administration and the development of Ardhisasa were discussed. Secondary data from digital media and publications from professional bodies provided more information regarding the public discussions revolving around Ardhisasa's development and implementation.

Ardhisasa's perpetual interstructural moment

Ardhisasa, as a digital land information management system, was initially envisioned as a one-stop shop designed to enable the State Department of Lands and Physical Planning to modernise the land administration system by improving efficiency and transparency in land transactions through a web-based platform (Kabubu and Wambui 2021). Kenya's significant surge in data volumes and its rapid population growth have led to notable repercussions on the efficiency and effectiveness of service delivery. The reliance on manual land information management systems within government offices has resulted in long queues and created an environment vulnerable to corruption (Kariuki et al. 2018). Ardhisasa aims to address these challenges by introducing a digital system that promises improved efficiency in land transactions and reduces corruption within the land market.

However, as we will explain in this section, this is a promise that neither Ardhisasa nor its predecessors have been able to keep. Ardhisasa is at best seen as incomplete, a work in progress, or, as described by a land valuer we interviewed, 'an ongoing conversation between, uh, the institution of surveyors in Kenya, who will present the affairs of the surveyors and state agents' (NA221110PV).

This ongoing conversation regarding the development of Ardhisasa has to be understood within a continuous cycle of innovation, incomplete implementation and reform that has characterised attempts to digitalise Kenya's land information management system, starting with the first NLIMS as part of the state's e-Government strategy introduced in 2004. Following this first attempt to implement a digital NLIMS, several others have followed (Table 5.1), each providing the incomplete, sometimes partially discontinued, foundation for the next iteration.

Hailed as the most ambitious attempt to digitalise Kenya's land administration system to date, in conversations with private-sector actors in the field of land administration and management Ardhisasa was continuously described as having great potential for reforming and streamlining the land administration system:

> Ardhisasa is supposed to facilitate all the large transactions and do away with manual processing of transactions. So you can search on it, get your land rate upon demand, a surveyor can carry out a subdivision and launch the whole process on that platform and complete from the comfort of our homes. That is the whole idea. We can carry out complex land transactions without really having to interact physically with the Ministry of Lands. It's a very noble project, I must say. Uh, but of course, it has had its own teething challenges. (K10112022-1)

After only a few months in operation, the digitisation of the paper-based land information management system came under fire from both public and private entities, raising questions about whether the current challenges are novel to digitisation or are only now visible as an exacerbation of already existing issues in Kenya's current land systems. Kenya's land sector is characterised by historical injustices due to colonisation, longstanding political tensions related to land, and the elite capture of resources and institutions (K'Akumu 2016; Boone et al. 2019). Aiming to address these problems, the period between 2009 and 2016 was characterised by drastic reforms in the land sector, most significantly through the new constitution enacted in 2010. Hence, the various iterations of digitalising the land information management system take place amid institutional and legislative reforms changing the administrative landscape in Kenya by decentralising land governance from the national state to the counties through the new Lands Act and the National Land Commission (Boone et al. 2019).

Some of the main challenges prevailing within the platform are related to the fact that with the introduction of Ardhisasa, the Ministry of Lands and Physical Planning is simultaneously working on the unification of the land title system across the country into a block system and geo-referencing governmental maps to create a national land cadastre. While officially separate projects, these are interlinked since the land records can only be included in Ardhisasa after they have been converted and geo-referenced. Delays in both of these processes have led to the incomplete inclusion of land records in the system, making it impossible to access

Table 5.1 Timeline of government-led initiatives to digitalise Kenya's land administration.

Year	Name	Description	Reason for succession
2004	e-Government	Milestone initiative to digitalise Kenya's government transactions, including in the Ministry of Lands and Physical Planning.	Continues to form the guiding strategy. After its adoption, several attempts have been made to digitise the land records, culminating in the current NLIMS.
2009	First Project on Land Administration in Kenya (PILAK I)	Funded by the Swedish International Development Agency with the objective of improving land administration as part of the development of the NLIMS to develop business and IT infrastructure, modernise the geodetic framework, and reform the parcel identification and land rent collection systems.	Only partially implemented due to continued challenges relating to the lack of security of paper records, parcel identification, land rent collection, geodetic reference framework, systematic conversion of Registered Land Act titles, staff capacity and awareness.
2013	Second Project on Land Administration in Kenya (PILAK II)	Address the challenges of PILAK I, and develop and implement a pilot integrated GIS-based NLIMS.	The system was to be implemented and funded through the Medium-Term Expenditure Framework budget for the period 2013–17, but failed due to lack of funds. Furthermore, the additional GIS component did not work. Unlike PILAK I, which was donor funded, PILAK II was to be government funded.

2014	Electronic Document Management System (EDMS)	Return to PILAK I and reorganise the land registries using an EDMS.	The system brought about huge backlogs and only dealt with land registries, whilst the Department of Surveys, which is the foundation of land administration, was left out.
2017	e-Citizen	Aimed at the digitalisation of all citizen–government transactions, including the payments of land rents and the statutory governmental charges in the land information management system.	Like the EDMS, e-Citizen was not GIS-based. The Department of Surveys was not incorporated, as the system dealt only with land registry backlogs in transactions due to lack of integration of the various departmental functions.
2021	Ardhisasa	An online platform that allows citizens and private-sector actors to interact with land information held and processes undertaken by the national government.	Currently being piloted in Nairobi County.

data and verify its status for the areas of Nairobi, which have yet to be converted to the block system, leading to a slowdown of land transactions. Further, for areas that have been converted to the block system, delays in receiving the results of the search were at times longer than through the previous manual system, elongating the land transaction period (Tarus and Wamae 2022).

Appearing before a senate committee, the former cabinet secretary for the Ministry of Lands and Physical Planning reiterated that the digitisation process of the Central Registry records (which holds documentation regarding land previously occupied by the white settlers) was to be completed in 2024 (*The Star* 2022). However, delays in these processes have severely impacted efficiency and service provision through Ardhisasa. One of the platform's core functions, the title authentication search – often referred to as an 'official search' – is unavailable for certain parcels of land that have yet to be digitised, effectively excluding sections of the city. This standard procedure is the starting point in the process of land valuation and is crucial for all land transactions. Those who have used the system to conduct searches say that the system is, at times, unable to generate results. Various private-sector actors in the land sector were frustrated by the lack of information flows to establish the authenticity of title documents: establish ownership, size, type of tenure, encumbrances, and the history of the parcel.

Contestation and negotiation

At the core of these developments are the private-sector actors influencing the design and development of Ardhisasa. While the private-sector actors we conversed with at times expressed frustration with the current circumstances, there seemed to be a general patience and willingness to collaborate in developing the platform from which they should ultimately benefit. This is not to say that the relationship between private-sector actors and the Ministry of Lands and Physical Planning is always constructive. One land surveyor explained that Ardhisasa had been developed top-down, with little regard for the expertise and needs of the professionals whose work practices will change due to its implementation. 'There have been meetings between, uh, our representative body, which is the institution of surveyors in Kenya, and the Minister of Land' (NA221110PV). However, he went on to explain that aside from these meetings, there had been minimal consultation with private-sector actors in the design phase. Another land surveyor explained that there was a willingness to cooperate and negotiate over the platform's features to ensure that its functionality meets the needs and matches the workflow of land professionals. Yet, in his experience, the 'developers' – referring to the programmers coding the digital platform – did not take their input seriously and failed to consider the usability of the platform for surveyors in the field.

In the development phase, private-sector actors had opportunities – albeit with varying degrees of influence – to shape the platform's development and voice their opinions in conventional settings such as stakeholder meetings, advisory taskforces and policy position papers. However, in the pilot phase, various professional bodies have resorted to public protest or aired their dissatisfaction in the media in an attempt to influence the platform's evolution. This has not been without impact.

As a platform, Ardhisasa officially grants access to documents and data by issuing verified accounts for private landowners and private-sector actors. However, at its initial implementation, Ardhisasa's architecture did not recognise land valuers as a relevant group alongside land surveyors and lawyers needing verifiable accounts and access to data on the platform. Following repeated protests and lobbying from the professional body of land valuers, a separate account option was created for these professionals. Similarly, following public complaints from the Law Society of Kenya regarding the fact that foreigners cannot register on the platform – one of the requirements for an account is a Kenyan ID number – the Ministry of Lands and Physical Planning has responded that it will work towards making this possible. It is in these interactions that Ardhisasa shows itself as a responsive platform, accommodating the needs and perspectives of private-sector actors as key users and stakeholders in the process.

Nonetheless, complaints about the changing workflows within the system have yet to be addressed. Under the paper-based land information management system, and in the digital predecessors to Ardhisasa, anybody with a stake in a registered piece of property could freely request and receive an official search from any land registry. This enabled individuals and organisations to access information on ownership and encumbrances of any registered land parcel whenever they chose, provided they paid the requisite statutory fees. Currently, in a break with custom, Ardhisasa requires that the property owner's permission is sought before conducting official searches. The negotiation about the workflow is, in essence, a discussion regarding the trade-off between transparency of information and the privacy of property owners, which is currently being resettled in practice and through debate in meetings, publicly on social media and through protest.

In addition to negotiating over the platform's architecture, Ardhisasa's introduction has forced private-sector actors in the land sector to adapt their practices. Adjusting the socio-material system by developing new relationships or deepening existing ones proves to be crucial to maintain information flows even if the digital platform does not produce the results required. According to one government planner we interviewed, some land professionals have managed to continue with land transactions because they have a liaison person whom they consult to facilitate the process. Actors who do not have an inside contact are forced to interact with a 'silent' platform that will not lead to quick results:

> The frustrating bit is that you don't know who to call. Because once you launch it on the platform, it's just wait and wait and wait unless you have an inside contact, and I have no idea what they're doing now on the inside. I'm not sure if they have access to the system and can see you applied for the search. I have no idea. You have to pay someone inside to do it for you, because you can't proceed with the valuation unless you have the searches. This also applies to the bankers and the lawyers who are dependent on the search document to effect any transaction. (K10112022-1)

Taken together, these expressions of frustration, experiences of transformation, and the need for adaptation among land professionals are illustrative of the developmental stage of Kenya's land administration system. It is in between paper and digital-based systems, in between the development and implementation phase, and in between previous innovations and future ones. As described earlier, the constant reinvention of digital platforms for land administration causes a seemingly never-ending transitional phase where one digital platform serves as a starting point for the next iteration, each different from the previous one through incremental innovations. However, none has been implemented fully and embedded in land administration practices.

It is through these series of ad hoc interactions and conversations, some conducted in formal settings while others happen behind closed doors or on the street, that Ardhisasa is taking shape. In this proactive form of engagement with the development of the digital platform, actors also find ways to overcome their precarious position and build more secure positions within the system that align with their previous roles in the paper-based system. For Ardhisasa, this means that although clear decisions on how to impact the development trajectory of the platform can be made, rules for interaction can also be withdrawn, adjusted or adapted.

Experimentation

Outside of the core of 'elite stakeholders' (Côté-Roy and Moser 2019), such as private-sector actors, civil society is peripheralised in Ardhisasa's development. One of the main concerns regarding the introduction of Ardhisasa has been the issue of digital illiteracy and inclusion. In Kenya, 17 per cent of the adult population has low literacy levels, and an even higher number of people, the majority women, are not fully digitally literate or have only limited access to digital technologies such as computers, smartphones and the internet (Koyama et al. 2021; UNESCO Institute for Statistics 2022). Moreover, while Swahili is the more commonly spoken language in Kenya, the platform is currently only available in English. As a result, a large portion of the Kenyan population will not have the capacity to fully engage with Ardhisasa in its current form. This has led to calls to maintain a paper-based system to fall back on in addition to the digital platform:

> So, we need to be inclusive ... and we have to look for another alternative, which gets us back to, takes us back to paper ... If you do not consider paper and then you end up leaving a section of people [out] ... The government is digitising land records, and that means you're going to digitise your land rights. (NA22110CS)

Nevertheless, what is less discussed is the effect of digitalisation on those for whom the paper bureaucracy also does not function. Spread across Nairobi are informalised settlements that house around 60 per cent of the urban population, assembled and expanded through unregularised land transactions and autoconstruction. Like so many other informalised settlements around the world, these neighbourhoods always seem to be many things at once. They are icons of the resilience and self-sufficiency of urban residents but also are in need of upgrading and development; they are examples of pioneering and bottom-up urbanism but also need state regulation and service provision; they are the stepping stone into the city while also being the embodiment of exclusion and marginalisation; they are vehicles for incremental consolidation while also being perpetually unfinished. More importantly, they are in constant transformation. They are spaces of continuous struggle to transition into more formalised urban assemblages, of which a crucial step is the regularisation of land titles and urban residency. The introduction of Ardhisasa means that land possession has to be registered on paper and in the digital land information management system. The absence of landholding documents in the official registers is automatically associated with non-ownership.

Yet, parallel to the formalised land title registration system, many other forms of land governance and transfers exist within Nairobi, which Ardhisasa does not cater to. In Nairobi's land market, a variety of alternative documents – share certificates, temporary occupation licences, allotment letters, certificates of tenure, subleases – are used to transfer land as proof of ownership in informal or semi-formal ways. While the state does not recognise these documents as confirmation of land ownership, they are generally accepted in property and land transactions as documenting unregistered housing and proof of continued occupancy. Hence, they are intermediary documents that play an important role in transitioning from informalised to formalised residency and ownership or, for example, determining who should be included in slum upgrading schemes and resettlement programmes. However, as they only establish a legitimate interest in the land rather than ownership, and are often not issued by a formal authority but are drafted amongst community members, they cannot be registered within Ardhisasa.

A lack of documentation is not only an issue for people living on squatted public or private land. Due in part to the backlog in registrations and the cumbersome bureaucratic process, many people do not follow through with registering transactions as a result of succession, inheritance or sub-divisions. As these transactions are often only processed informally, without officially changing the land title, disputes over land ownership and land use claims in Nairobi are prevalent. With the digitisation of land records, property owners are confronted with a lack of both administrative oversight and formal paper-based and digital registration of land titles.

Multiple community-led groups and civil society organisations in Nairobi and Kenya at large have aimed to raise issues related to digital inclusion and the registration of people living in different degrees of informality through various means, such as lobbying, organising multi-stakeholder workshops, and solicited and unsolicited advice. Most notably, there have been initiatives to take matters into their own hands – to map and register irregular settlements by and for the community. They use various digital and in-person means to do this, overcoming the constraints of the existing digital platforms. As a civil society organiser stated, '[g]overnment has been absent for a long period of time. And when government is absent, then people innovate' (NA230317I011).

In this case, the innovation she was referring to was the introduction of an alternative land registration system – the social tenure domain model (STDM) – where the starting point is the relationship between an individual and the space they occupy rather than the documents they possess. The STDM functions as a tool for land administration that breaks down the complexities that exist within irregular settlements. It allows us to identify the communities and individuals living on squatted land, and define and establish their relationship with space and their role within the land market. For example, individuals can be registered as tenants, structure owners or landlords. Moreover, exploring the opportunities and boundaries provided by the Community Land Act, the STDM facilitates the registration of communal lands in informalised urban neighbourhoods (Otieno et al. 2017). While the Community Land Act was introduced in 2016 to formalise rural community lands, it provides a comprehensive definition of community, opening pathways for urban collectives, living on the same land, sharing customs, culture or language, to pursue the registration of the territory they occupy as community land.

The primary motivation for introducing the STDM in Nairobi is to expand the formal categories of land tenure to include the relationships with land and occupancy that have emerged through informality. This challenges the structures determining 'legitimate' land tenure and works towards decriminalising the autoconstruction of housing and basic services such as water and electricity in informalised settlements, hoping to reduce forced evictions and provide security of tenure.

On a more profound level, the experiments with community mapping, the STDM and registering irregular urban settlements under the Community Land Act challenge the assumptions in the architecture of Ardhisasa that value private ownership and public administration structures over communal ownership or collective land administration and management. Where private-sector actors such as land valuers, lawyers and surveyors are making progress in influencing Ardhisasa to become a faster, more accessible and more efficient platform, the community organisations pushing these alternatives want something far more radical. They want to introduce a non-capitalist land administration system that prioritises communal over private ownership and appreciates the African socio-cultural dimensions of land. In discussing the tensions between the formal systems of the state and the informal and semi-formal documentation of communities that want to become formalised, a community organiser explained the challenges the assumptions built into the NLIMS:

> So how do we legitimise their innovations to form a basis of policy development, for instance, in the land sector? There are layers of rights that people have in the informal settlements. They are user rights. They are access rights. They are occupancy rights … That has been the contestation between the formal Kenya and the informal Kenya that forms the majority. And it will lead us to our conversation in the National Land Information Management System … The National Land Information Management System, as it was built, does not appreciate the people's interpretation and the cultural interpretation of land as property in the African context. (NA230317I011)

The interstructural moment of the land administration system – a time-space and praxis in which the categories are still malleable – allows for these types of innovations, tensions and discussions to emerge. It provides an entry point for civil society organisations to introduce alternative approaches that question the basic premises of the platform. Yet, we notice that while private-sector actors have exerted significant influence on the development of Ardhisasa as a platform, civil society organisations and community representatives have had little real influence to date. Ardhisasa is constructed through politics, knowledge and actors, and emerges from a market-driven governance approach. The objectives of the land professionals gel well within the neoliberal imaginary of the digital state that Kenya is trying to establish.

On the contrary, community groups and civil society organisations propose a radical alternative that challenges the platform's core objectives and is therefore not included in the conversation. This illustrates the peripheralisation of these groups in their attempts to be recognised as important actors in the digitisation of land records. As a result, people living in various degrees of informalised infrastructure space in Nairobi continue to be blank spots in the platform. They operate from the informational peripheries, socio-spatially distant and outside the information systems governing land. Yet, as Kimari and Ernstson write, 'rather than seeing the innovative infrastructural and incremental practices of the marginalised as making do, or filling the gaps, we can see them as crucial sites to rethink spatial distributions of power' (Kimari and Ernstson 2020, 9). Creating new forms of information and new ways to record land and property fall within these modes of living between defiance and compliance (Easterling 2016).

Conclusion: spatialising the informational peripheries of Ardhisasa

Looking critically at the negotiations over the platform and whose voices are represented or not in the digital land administration system reveals how notions about legitimate tenure, ownership and land governance are embedded in the platform's architecture. In this chapter, we specifically did not focus on the formal institutions and how they have built Ardhisasa. Instead, we analysed the platform's emergence through the conversations, contributions and conflicts with private-sector and societal actors that Kenya's land administration relies on. The various cycles of design, implementation and abandonment of policies and digitalisation efforts have turned what promised to be an all-encompassing e-Government strategy into an enduring process of contestation, negotiation and experimentation, which never really seems to reach completion. Whilst Ardhisasa, like many other state-led projects in Kenya, tries to present a smooth transition and a coherent set of practices for land information management and service delivery, the reality proves to be more precarious (Lesutis 2022a). In a context as complex and fraught as land administration in Kenya, there are multiple objectives and rhetoric that continually bump heads with each other, sometimes through negotiation and experimentation and sometimes through contestation, leading to an open-endedness in the process of platformisation.

Our analysis highlights how the development of digital platforms is situational, power-laden, and a politically ambiguous praxis. Ardhisasa reflects the very ambiguities that gave rise to it in the first place. In its becoming, Ardhisasa exists in between the ambitions of Kenya's authorities to reduce bureaucracy and corruption and the social and political imperatives of a highly unequal state and urban society. These often conflicting foundations allowed negotiation over and experimentation with the rationale and function of the platform and its uses. We note that Ardhisasa is remarkable in its flexibility in incorporating feedback from private-sector actors. The digitalisation of public services through diverse infrastructural interventions, such as land information management systems, provides an opportunity for private-sector actors to reconfigure inherently political socio-technical relations and consolidate their position within the system. The platform is expanded or adjusted through negotiations with land professionals without challenging its underlying principles. At the same time, Ardhisasa has proven to be resistant to the efforts of civil society organisations that challenge the assumptions in its architecture and, by extension, the neoliberal politics of the platform. These actors remain in the informational periphery, outside the geographical spaces and digital structures of power (Datta 2024). This mirrors the dynamics described by Kimari and Ernstson (2020), where large-scale infrastructural projects in Kenya reproduce the landscapes of empire and capitalist production under the banner of 'development' or 'modernity', neglecting citizens' actions that envision more just, communal or non-capitalist futures.

As a result, Ardhisasa produces the informational periphery. The development of the land information management system is inherently situated in space and the inequalities existing with the city and land sector. Analysing the development of Ardhisasa in relation to the experimental alternatives proposed by civil society organisations in Nairobi casts a light on how the platform's architecture structures the urban landscape in profound ways. The continuous development of the digital platform and the fact that it is incomplete both in its implementation (not all data is included) and in its scope (not all land transactions can become digitalised) contribute to peripheral existence in informalised settlements. Hence, considering the relationship between the digital platform and the informalised city, we note the futility of approaching 'land' and 'digital' as separate fields of urban governance; rather we argue for a critical analysis of digital-land relations and the territorialisation of information infrastructures. Through the perpetual becoming of the LIMS implemented in Kenya, the spatial periphery overlaps with the informational periphery, making informalised communities not only geographically but also informationally distant from the centres of regulation, decision making and power. In many ways, this is a silent process of further exclusion of already disenfranchised communities living in conditions of semi-formality and informality by increasing uncertainty. As Ardhisasa is currently piloted in Nairobi, we have focused our analysis on its impact on informalised urban settlements. However, as Ardhisasa completes its anticipated roll-out nationally, it will become important to consider the future impact on Kenya's rural and indigenous communities, which are materially, politically and informationally underserved by the state's institutions.

Acknowledgements

An earlier version of this chapter was published as F. I. Hoefsloot and C. Gateri, Contestation, negotiation, and experimentation: The liminality of land administration platforms in Kenya (2024).

Environment and Planning D: Society and space 42(5–6). https://doi.org/10.1177/02637758241254943.

Funding for this research was provided by European Research Council Advanced Grant 'Regional Futures: The territorial politics of digitalisation-as-urbanisation in the Global South' (ID: 101019318) awarded to Ayona Datta. We are grateful to Ayona Datta and Dennis Mbugua Muthama for their help in conducting the interviews and for feedback on earlier versions of this paper. We also want to thank all the land professionals and civil society organisers who participated in this research and shared their knowledge and time with us.

References

Anand, N., Gupta, A. and Appel, H. (eds). 2018. *The Promise of Infrastructure*. Durham, NC: Duke University Press.

Artioli, F. 2018. *Digital Platforms and Cities: A literature review for urban research*. SciencesPo Working Paper No. 01/2018. Accessed February 2025. https://sciencespo.hal.science/hal-02385137.

Barns, S. 2018. Smart cities and urban data platforms: Designing interfaces for smart governance. *City, Culture and Society* 12: 5–12. https://doi.org/10.1016/j.ccs.2017.09.006.

Boone, C., Dyzenhaus, A., Manji, A., Gateri, C. W., Ouma, S., Owino, J. K., Gargule, A. and Klopp, J. M. 2019. Land law reform in Kenya: Devolution, veto players, and the limits of an institutional fix. *African Affairs* 118(471): 215–37. https://doi.org/10.1093/afraf/ady053.

Bowker, G. C. and Star, S. L. 2000. *Sorting Things Out: Classification and its consequences*. Cambridge, MA: MIT Press.

Calzada, I. 2018. (Smart) citizens from data providers to decision-makers? The case study of Barcelona. *Sustainability* 10(9). https://doi.org/10.3390/su10093252.

Côté-Roy, L. and Moser, S. 2019. 'Does Africa not deserve shiny new cities?' The power of seductive rhetoric around new cities in Africa. *Urban Studies* 56(12): 2391–407. https://doi.org/10.1177/0042098018793032.

Cowan, T. 2021. Uncertain grounds: Cartographic negotiation and digitized property on the urban frontier. *International Journal of Urban and Regional Research* 45(3): 442–57. https://doi.org/10.1111/1468-2427.13016.

Datta, A. 2023. The digitalising state: Governing digitalisation-as-urbanisation in the global south. *Progress in Human Geography* 47(1): 141–59. https://doi.org/10.1177/03091325221141798.

Datta, A. 2024. The informational periphery: Territory, logistics and people in the margins of a digital age. *Asian Geographer* 41(2): 125–42. https://doi.org/10.1080/10225706.2023.2253233.

Easterling, K. 2016. *Extrastatecraft: The power of infrastructure space*. London: Verso.

Elbanna, A. and Idowu, A. 2022. Crowdwork, digital liminality and the enactment of culturally recognised alternatives to Western precarity: Beyond epistemological *terra nullius*. *European Journal of Information Systems* 31(1): 128–44. https://doi.org/10.1080/0960085X.2021.1981779.

Guma, P. K. 2020. Incompleteness of urban infrastructures in transition: Scenarios from the mobile age in Nairobi. *Social Studies of Science* 50(5): 728–50. https://doi.org/10.1177/0306312720927088.

Heeks, R., Gomez-Morantes, J. E., Graham, M., Howson, K., Mungai, P., Nicholson, B. and Van Belle, J-P. 2021. Digital platforms and institutional voids in developing countries: The case of ride-hailing markets. *World Development* 145. https://doi.org/10.1016/j.worlddev.2021.105528.

Hoefsloot, F. I., Richter, C., Martínez, J. and Pfeffer, K. 2022. The datafication of water infrastructure and its implications for (il)legible water consumers. *Urban Geography* 44(4): 729–51. https://doi.org/10.1080/02723638.2021.2019499.

Jordan, Tim. 2015. *Information Politics*. London: Pluto Press.

Kabubu, J. and Wambui, C. 2021. Ardhisasa – The ABCs. MMAN Advocates, 17 November. Accessed February 2025. https://mman.co.ke/content/ardhisasa—-abcs.

K'Akumu, O. A. 2016. Mapping stakeholder positions in the Kenyan land reform process. *International Journal of Technology Management & Sustainable Development* 15(1): 15–36. https://doi.org/10.1386/tmsd.15.1.15_1.

Kariuki, J. W., Karugu, D. W. N. and Opiyo, M. M. O. 2018. Challenges facing digitization projects in Kenya: Case of implementation of National Land Information Management System. *International Journal of Technology and Systems* 3(1): 23–42.

Kimari, W. and Ernstson, H. 2020. Imperial remains and imperial invitations: Centering race within the contemporary large-scale infrastructures of East Africa. *Antipode* 52(3): 825–46. https://doi.org/10.1111/anti.12623.

Kitchin, R., Maalsen, S. and McArdle, G. 2016. The praxis and politics of building urban dashboards. *Geoforum* 77: 93–101. https://doi.org/10.1016/j.geoforum.2016.10.006.

Kitchin, R., Coletta, C., Evans, L., Heaphy, L. and MacDonncha, D. 2017. Smart cities, epistemic communities, advocacy coalitions and the 'last mile' problem. *it – Information Technology* 59(6): 275–84. https://doi.org/10.1515/itit-2017-0004.

Koyama, N., Totapally, S., Goyal, S., Sondereffer, P., Rao, P. and Gosselt, J. 2021. Kenya's digital economy: A people's perspective. Dalberg. Accessed February 2025. https://dalberg.com/wp-content/uploads/2021/11/Kenyas_Digital_Ecomony_Full_report_Aug_2021-1-1.pdf.

Lancione, M. and Simone, A. 2021. Dwelling in liminalities, thinking beyond inhabitation. *Environment and Planning D: Society and space* 39(6): 969–75. https://doi.org/10.1177/02637758211062283.

Lesutis, G. 2022a. Disquieting ambivalence of mega-infrastructures: Kenya's Standard Gauge Railway as spectacle and ruination. *Environment and Planning D: Society and space* 40(5): 941–60. https://doi.org/10.1177/02637758221125475.

Lesutis, G. 2022b. Politics of disavowal: Megaprojects, infrastructural biopolitics, disavowed subjects. *Annals of the American Association of Geographers* 112(8): 2436–51. https://doi.org/10.1080/24694452.2022.2062292.

Leszczynski, A. 2020. Glitchy vignettes of platform urbanism. *Environment and Planning D: Society and space* 38(2): 189–208. https://doi.org/10.1177/0263775819878721.

Li, T. M. 2007. Introduction: The will to improve. In *The Will to Improve: Governmentality, development, and the practice of politics*, edited by T. Murray Li, 1–30. Durham, NC: Duke University Press.

Lutzoni, L. 2016. In-formalised urban space design: Rethinking the relationship between formal and informal. *City, Territory and Architecture* 3(1): 20. https://doi.org/10.1186/s40410-016-0046-9.

Manji, A. 2020. *The Struggle for Land & Justice in Kenya*. Martlesham: Boydell & Brewer.

McConnell, F. 2017. Liminal geopolitics: The subjectivity and spatiality of diplomacy at the margins. *Transactions of the Institute of British Geographers* 42(1): 139–52. https://doi.org/10.1111/tran.12156.

Müller, M. and Trubina, E. 2020. Improvising urban spaces, inhabiting the in-between. *Environment and Planning D: Society and space* 38(4): 664–81. https://doi.org/10.1177/0263775820922235.

Odendaal, N. 2023. *Disrupted Urbanism: Situated smart initiatives in African cities*. Bristol: Bristol University Press.

Omura, K., Otsuki, G. J., Satsuka, S. and Morita, A. 2019. *The World Multiple: The quotidian politics of knowing and generating entangled worlds*. Abingdon: Routledge.

O'Reilly, T. 2011. Government as a platform. *Innovations: Technology, governance, globalization* 6(1): 13–40. https://doi.org/10.1162/INOV_a_00056.

Orlikowski, W. J. and Scott, S. V. 2021. Liminal innovation in practice: Understanding the reconfiguration of digital work in crisis. *Information and Organization* 31(1). https://doi.org/10.1016/j.infoandorg.2021.100336.

Otieno, E., Lengoiboni, M., Bennett, R. M. and Ayugi, S. 2017. Analysing the impact of social tenure domain model (STDM) on tenure security in an informal settlement. In *Proceedings of the EALAN Conference*, Nairobi, Kenya, 26–27 July, 64–86. Accessed February 2025. http://land.igad.int/index.php/documents-1/improving-land-governance/1525-promoting-land-administration-and-governance-for-sustainable-development-in-eastern-africa/file.

Pelizza, A. 2017. Disciplining change, displacing frictions: Two structural dimensions of digital circulation across land registry database integration. *TECNOSCIENZA: Italian journal of science & technology studies* 7(2): 35–60. https://doi.org/10.6092/issn.2038-3460/17326.

Poell, T., Nieborg, D. and van Dijck, J. 2019. Platformisation. *Internet Policy Review* 8(4). https://doi.org/10.14763/2019.4.1425.

Senshaw, D. and Twinomurinzi, H. 2022. Innovating with government digital platforms in low-income countries: The dynamic capabilities of Woredas in Ethiopia. *Journal of Science and Technology Policy Management* 13(4): 812–36. https://doi.org/10.1108/JSTPM-05-2020-0081.

Siles, I., Gómez-Cruz, E. and Ricaurte, P. 2023. Toward a popular theory of algorithms. *Popular Communication* 21(1): 57–70. https://doi.org/10.1080/15405702.2022.2103140.

The Star. 2022. Digitisation of land records to be completed in 2024 – Karoney. Accessed February 2025. https://www.the-star.co.ke/news/2022-03-03-digitisation-of-land-records-to-be-completed-in-2024--karoney.

Tarus, E. and Wamae, P. 2022. Land records digitization and service delivery in the Ministry of Lands in Kenya. *International Journal of Current Aspects* 6(3): 59–69. https://doi.org/10.35942/ijcab.v6i3.278.

Taylor, L. and Richter, C. 2017. The power of smart solutions: Knowledge, citizenship, and the datafication of Bangalore's water supply. *Television and New Media* 18(8): 721–33. https://doi.org/10.1177/1527476417690028.

Turner, V. W. 1967. Betwixt and between: The liminal period in *rites de passage*. In *The Forest of Symbols: Aspects of Ndembu ritual*, 93–111. Ithaca, NY: Cornell University Press.

UNESCO Institute for Statistics. 2022. Literacy rate, adult total (% of people ages 15 and above) – Kenya. Accessed February 2025. https://data.worldbank.org/indicator/SE.ADT.LITR.ZS?locations=KE.

van Dijck, J. 2021. Seeing the forest for the trees: Visualizing platformization and its governance. *New Media & Society* 23(9): 2801–19. https://doi.org/10.1177/1461444820940293.

Vanolo, A. 2016. Is there anybody out there? The place and role of citizens in tomorrow's smart cities. *Futures* 82: 26–36. https://doi.org/10.1016/j.futures.2016.05.010.

Theme II

Territorialising the informational periphery

6

The logistical periphery: territory, land and people in the margins of a digital age

Ayona Datta

In April 2022, I was in a taxi heading towards Bhiwandi, a small town located on the north-eastern periphery of the Mumbai Metropolitan Region. Bhiwandi has been known for a very long time as the 'Manchester of India': its power loom industries have shaped the textile economy of the region since the early nineteenth century. The road from Mumbai to Bhiwandi passes through a landscape of extremes – lush green farmland, dense high-rise housing developments, office complexes, chemical factories, busy motorways, roadside service stations and perpetual traffic jams. The journey from Mumbai to Bhiwandi can take up to five hours during the monsoons.

This is a heavily urbanised industrial corridor, and yet as you reach Bhiwandi you realise that this is a place like no other town in Mumbai's fringes. Unlike other satellite towns of Mumbai, such as Kalyan or Thane, Bhiwandi feels out of place within Mumbai's striking global reach. As a manufacturing town, almost half (49 per cent) of Bhiwandi's one million population live in slums and informal settlements. It exhibits all the outward signs of a developmental crisis – it has a low literacy rate (79.48 per cent), a low sex ratio (709 females:1,000 males), mainly because of the migrant labour, and a majority (56 per cent) Muslim population, with high indices of deprivation and with poor access to urban basic services such as water, sanitation, energy and network connectivity. It also has very high digital inequalities among young Muslim women, elderly and migrant workers in terms of access and capacity to use mobile phones and digital platforms (Shaban et al. 2023).

Figure 6.1 A Bhiwandi power loom factory. Photograph: Rohit Madan, 2022.

One can hear Bhiwandi before reaching it. By this I do not just mean the noise of the traffic and the people on the streets. Rather, as one approaches the city, particularly at night or in the early hours of the morning when the traffic is slower, one can hear the constant hum of the power looms (Figure 6.1) – the continuous click clack of the warp and weft which are woven into some of the best cotton cloths in India

Figure 6.2 A Bhiwandi logistics park. Photograph: Ayona Datta, 2022.

– supplying raw material for well-known textile hubs in other parts of India such as Gujarat and Rajasthan. These power looms do not stop – ever. They slow down during the Friday prayers, but the constant hum gives an auditory identity to the city – a working home-based industrial town made and run by migrant labour from across India. As one loom worker said to us, 'You can come to Bhiwandi in the morning, and you will find work by evening.'

But just outside Bhiwandi, on the Mumbai–Agra Highway, one begins to see a new kind of territorialisation that has nothing to do with Bhiwandi's history. Rather this is a landscape of global logistics manifesting in the architecture of India's digital age. Mile upon mile of the highway that stretches from Bhiwandi is dominated by mega-scale warehouses (Figures 6.2, 6.3). Indeed, Bhiwandi is now India's largest e-commerce and logistics hub, with companies like Amazon, Flipcart and Nyka holding their goods here. This is the paradox of India's digital urban age – while Bhiwandi has long been bypassed by Mumbai's regional development initiatives, the advancement of digital infrastructures and technologies in the region sustains some of the most sophisticated

Figure 6.3 Inside a warehouse in Bhiwandi. Photograph: Rohit Madan, 2022.

logistics facilities in Bhiwandi. This, then, is the periphery in a digital age where territory, logistics and people come together in contradictory ways to produce a networked marginality.

Fuelled by a ready supply of large agricultural land parcels around Bhiwandi, logistics now defines the temporality of the city. Late evening and early morning drives in and out of Bhiwandi mean navigating highways full of HGVs and lorries, many of which frequently break down or crash along the roadside. A whole bevy of roadside eateries – or *dhabas* – has emerged along the highway, fuelling a vibrant local nightlife that caters to its youth and businessmen. Bhiwandi's complex transformation raises the question – what constitutes the urban periphery when logistics stretches the very idea of the urban beyond its territorial boundaries? When logistical territories reconfigure marginalities and informational flows across a region, what constitutes the informational periphery in a digital age?

In this chapter I will examine how multiple marginalities across land, infrastructure and people materialise as the 'informational periphery' (Datta 2024). Here I take the definition offered by Zins to distinguish between data and information:

> Data are the basic individual items of numeric or other information, garnered through observation; but in themselves, without context, they are devoid of information. Information is that which is conveyed, and possibly amenable to analysis and interpretation, through data and the context in which the data are assembled. (Zins 2007, 479)

Information is more than data – it is relational and includes both digital and analogue forms of communication; it includes narratives, stories, deals and dialogues between people, through face-to-face, paper-based or digitally mediated communication infrastructures. Information refers to contextually relevant data to which a certain logic has been applied for it to carry meaning and serve a purpose (Jordan 2015). Here, I follow the introduction to this book to note that the periphery in a digital age needs to be understood through the lens of information – encompassing both its politics and its flows through people, spaces and information technologies. Information politics, as Jordan notes, is the power vested in the flows, circuits and networks of information and the laws, policies and governance systems that produce its unevenness across spaces and territories (Jordan 2015). Informational flows are an extension of these politics in the ways that particular geographies and people are bypassed or actively excluded from a global logistical network through laws, policies and infrastructural redlining. In particular, I am interested in how information is simultaneously social, political and technological – producing specific geographical inequalities and exclusions through logistics.

I argue that Bhiwandi captures a territorial politics of the 'informational peripheries' (Datta 2024) that can offer us a different vantage point for understanding the spaces of exclusion and fragmentation created in the margins of metropolitan regions in the Global South. The informational periphery extends the idea of a periphery beyond geography to informational space. It attends to diverse and heterogeneous forms of digitally mediated urbanisation that are taking shape in metropolitan regions. Instead of considering the periphery as a primarily geographical metaphor, I approach informational peripheries as both geographically and digitally distant space that is entangled with regimes of territory, logistics and people. I propose that the empirical and theoretical development of informational peripheries in the context of logistics can provide an important vantage point for interrogating the political and technological relations that are at the heart of the emergence of urban peripheries in the 'digitalising state' (Datta 2023).

In this chapter, I will make three key arguments. First, moving beyond earlier conceptualisations of the 'infrastructural turn', I argue that the informational periphery should be seen through the lens of a 'logistics turn' (Keith and de Souza Santos 2021; Mutter 2023; Cowen 2014; Cavalier 2016; Lyster 2016). Logistics focuses exclusively on mobility, circularity and speed, which in turn depend on the capability of information to move smoothly and to remain accessible on demand. Cavalier calls this *logistification* (Cavalier 2016), which (like industrialisation) is a process of transformation that produces a frenzied rush of information through space and time to speed up the movement of material things. Cavalier notes that logistification may appear to be embedded nowhere, but it is a deeply territorial process. Seen through this lens of logistification, the periphery emerges as an informational space inherently tied to the territorialisation of global e-commerce (in the form of warehouses and logistics parks) that circulates capital, goods, labour and information precisely because of the ways in which they are networked through the digital.

My final argument is that transformations in the informational periphery deploy extractive technologies of data, labour and resources, which can be seen as forms of 'settler colonialism' (Yiftachel 2009). Logistics is enabled by flows of people – their embodied labours shape the ways global e-commerce sticks in and flows through these places by pushing labouring bodies within its shadows. However, settler colonialism in the context of Bhiwandi is not the long-term occupation of land by colonisers, but rather a more complex process where colonialism unfolds by proxy. I argue that forms of settler colonialism are evident in the ways that global logistics settles in Bhiwandi as warehouses, and the local networks of caste-ist, ethnic and religious connections to land and its accumulation enable this stickiness.

Informational peripheries produced in/through logistical spaces

For Simone, peripheries have a double status – as a 'space of insufficiency and incompletion' (Simone 2022, 40) – and although 'peripheries' refers to the outskirts of cities in common usage, Simone urges us to rethink cities in the Global South as also located within the peripheries of urban theory. Simone's development of this concept as the 'surrounds' – 'a relational location rather than a geographic one' (Simone 2022, 6) – means spaces that are

> heterotopic, exceptional, intensely specific, hidden in plain sight, prefigurative, or dissolute. In all instances the surrounds are infrastructural in that they entail the possibilities within any event, situation, setting or project for something incomputable and unanticipated to take (its) place. (Simone 2022, 5)

In this form of conceptualisation, the periphery is conceived as temporal and fluid, with a relational connection to the city and infrastructures. Although Simone does not connect this description with digital infrastructures per se, the periphery through his lens can be seen as a continuously shifting terrain of lived experience within the nodes of infrastructure and information lockdowns. The periphery is now located simultaneously across the centre and edge, across material and digital worlds. The periphery is the new frontier of urban extensions.

I extend this conceptualisation of the networked periphery by incorporating a logistical lens to understand the shifting and relational nature of urbanisation in a digital age. This follows on from the rich scholarship on the 'infrastructural turn', which has highlighted the ways in which infrastructures produce new spaces of power in the urban and peri-urban margins (Easterling 2016; Furlong 2020; Kanai and Schindler 2018; Du 2019). Indeed, Easterling argues that infrastructural space forms the bedrock of how the state governs through 'extrastatecraft' (Easterling 2016). However, an infrastructural lens seems inadequate to explain Bhiwandi's territorial expansion through the mushrooming of warehouses, which on the outside has been driven by the circulation of global logistics networks, and from the inside by the speed of informational exchange enabled by a local population divided along religious, caste and ethnic lines. A wider understanding of its transformation requires a closer look at what has recently been called a 'logistics turn' in urbanisation (Keith and de Souza Santos 2021; Mutter 2023; Cuppini and Frapporti 2018; Lyster 2016; Cavalier 2016). Keith and de Souza Santos argue that 'logistics emerges as a category of analysis as the combination of infrastructures, networks and urban speed' (Keith and de Souza Santos 2021, 168). Although logistics has a genealogy in the war machine, particularly in its emphasis on speed and connectivity of military operations, it is also deeply entrenched in the capitalist flows of goods, commodities, information and people, circulating and moving them great distances at tremendous speed (Chua et al. 2018). As Chua notes, logistics is 'animated by both the art of war and the science of business' (Chua 2017, 167) in space and labour.

A logistics turn in the periphery thus emerges from the deep relationship between information, territory and people. In her important work *The Deadly Life of Logistics*, Deborah Cowen (2014) argues that logistics is at the heart of contemporary political economies of globalisation. Logistics, as Lyster notes, draws together places 'anywhere, nowhere and everywhere' (Lyster 2016, 18). But logistics engages with places in specific ways, as Cavalier argues: 'Rather than seeking density, logistics aspires to coverage' and therefore is 'an agent in the transformation of territory' (Cavalier 2016, 6). Logistics includes infrastructure, but it extends beyond infrastructural space. It brings together 'informational systems, physical systems and mediating systems' (Cavalier 2016, 51), thus moving beyond infrastructural fluidity to territorial fluidity. But in order to circulate anywhere, logistics also needs to materialise as holding spaces somewhere – in territories where it can remain ready, on-demand, and available to move anytime, anywhere at speed. Logistics is mobilised through warehouses, roads, ports, trucks, wires and cables which carve out territories for time-space compression. Logistics is sticky as it descends in places that are located at the nodes of a vast global network which seeks to establish maximum coverage across territories. Logistics leverages the transformation of territory even as it appears to become invisible in terms of its territorial reach everywhere.

This lens, which sees logistics as a globally circulating territorial network, brings in a further understanding of logistics as an informational-territorial construct. Critical geographers have argued that logistics is dispersed across global networks, but its reach and significance are concentrated in highly local and territorial ways (Chua et al. 2018). This impacts on the ways in which territory is reorganised and recast in the service of a global logistics. Logistics requires infrastructure and labour, which are to be found in metropolitan areas. This makes logistics a key feature of urbanisation and metropolitanisation. As Cidell notes, '[t]he decisions being made today about land use within those territories regarding the location and appearance of the facilities that constitute the global logistics network will have impacts on the landscape, resource use, and economic and social geography of metropolitan areas for decades to come' (Cidell 2011, 832). Logistics depends on speed, which in return depends on the capability of these flows to move smoothly as well as to stick around within bounded territories for when required on-demand.

This feature of stickiness during mobility is always in need of territory that is ready for transformation in the service of logistics. As Cavalier argues, 'logistical territory is fungible' (Cavalier 2016, 35) – in other words, it is based on a general set of characteristics. The fungibility of territory depends upon its positioning within the informational periphery. On the one hand, this territory is positioned at the intersections of high-speed transport and digital infrastructure flows, and on the other, it lies at the margins of social and informational access. As Chua notes (drawing upon Mike Davis's work), logistics is the capital's underbelly, as it is this 'Logistical City' (Chua 2021) that lies at the heart of the flows of goods even as the infrastructure of the city moves 'stuff' further and further away from its highly infrastructuralised centre. Logistics, as Chua notes, is the driver of mega-infrastructure and transport projects as speed and connectivity are essential for making the logistics sector more efficient. Logistics, then, is at the heart of urbanisation even though it is kept well hidden from and peripheral to most conceptualisations of the 'urban'.

There is another way to understand the impact of logistics on urbanisation. Chua's important work on logistics highlights the racialisation of bodies and architecture in the global flows of goods. 'Logistics distribute vulnerabilities,' as Chua notes (2021, 262), since the insatiable need for labour in support of the mobility and circulability of goods directs vulnerable populations and social groups along its infrastructural pathways. This links logistics with the uneven nature of urbanisation and marginalisation in the urban peripheries. But information also follows logistics, as the need to maintain real-time information about mobility and circulation of goods is foundational in the logistics sector, and this information is core to its sustainability and growth. In the logistics sector, information is highly striated along specific data corridors, specialised in the use of technologies for real-time communication, and dependent on the specific data on global supply/demand. For example, information on procurement, packaging, availability, supply, delivery of goods and the labour that makes this possible would be considered a priority, and therefore public resources and infrastructure would be allocated to achieve this goal. Information seen as extraneous to logistics would be peripheralised – often this relates to conditions of work and welfare in the logistics sector. Thus, as logistics reconfigures marginalities, it also pushes those facing increasing marginalisation more and more into the informational peripheries, as they are made invisible in the logistical information. Vulnerable groups are also unable to access meaningful information, while being available as precarious labour for efficient flows of goods.

This link between logistics and vulnerabilities is not entirely new. Several scholars now claim that logistics is rooted in the long histories of colonialism, imperial conquest and ecological resource destruction (Chua et al. 2018; Cowen 2014; Zeiderman 2021). Cuppini and Frapporti further argue that colonisation was the earliest form of logistics, with the slave trade forming a network of global supply, demand and flow of human bodies reduced to bare life that then enabled access to and extraction from territories. They note:

> [M]odern logistics is not just about how to transport large amounts of commodities or information or energy, or even how to move these efficiently, but also about the sociopathic demand for access: topographical, jurisdictional, but as importantly bodily and social access. (Cuppini and Frapporti 2018, 96)

Cuppini and Frapporti's colonial genealogy of logistics can be extended to examine warehouses as a form of colonial extractivism over land and labour. Warehouses represent the extractive nature of global logistics, but its architecture is built upon historically constructed relationships with land, labour and social networks of power in local contexts.

> Cities in settler-colonial contexts … occupy a paradoxical kind of site in relationships between colonizer and colonized. They occupy Indigenous lands and form a central component of the settler society, yet at the same time render Indigeneity profoundly out of place. The settler city is often portrayed as a symbol of a 'new world', a space of liberalism and democracy, a hub of globalization, a magnet for international migration, or a center of investment and corporate power – all dominant discourses that conceal their ongoing colonial nature. (Porter and Yiftachel 2019, 177)

At first glance, logistics is not quite the same as settler colonialism. Unlike colonialism, settler colonialism is defined by a desire to stay on the land for good, creating new opportunities to accumulate wealth. The settler nature of Israeli settlements, for example, does not seem quite the same as warehouses, which do not 'house' settlers per se. Indeed, warehouses are there precisely for storing goods, which have to keep moving fast, on demand, in real time, to urban consumers across the country. However, at the same time, in Bhiwandi these warehouses are here to stay, as their status as regularised buildings housing some of the most highly robotised and securitised warehouses in the country and as the future of the region (highlighted in the Regional Development Plan) means that global e-commerce (Amazon, Flipcart, Nykaa, Big Basket and others) has settled there by proxy. On the books, e-commerce companies do not own, rent or hire anything – land or buildings or employees. They are located within an information gap across different regulatory mechanisms that make warehouses economically viable for different stakeholders.

The logistics sector channels information flows and infrastructures towards an urban extractivism in the peripheries that begins to reflect forms of settler colonialism. Logistics in this context is not just an extractive relationship between coloniser (global e-commerce companies) and colonised (local communities), rather it is an internalisation of generations of caste, religious and ethnic power within local communities that enables global capital to settle by proxy in the warehouses. Here settler colonialism is hidden in plain sight as the warehouses represent a permanent temporariness in its materiality – regulatory policies that keep it outside formal planning mechanisms, as well as the prefabricated nature of these vast spaces that enable speedy assembly and dismantling on demand. This arrangement claims warehouses as short-term holding sites rather than long-term settlements.

As Mutter notes, logistics 'leverages value not only from the city's peripheries, but from its gaps: from the (supposedly) disused, latent or otherwise lacking' (Mutter 2023, 162). This does not mean that actually existing places such as Bhiwandi are disused or lacking in innovation or enterprise, rather the narrative of 'lack' denies Bhiwandi's capability to produce alternative urban futures in its existing power loom industry. By this narrative, the geographies of logistics can make themselves stick faster in the urban and metropolitan peripheries, expand their infrastructural reach and access to territories and the people therein, and present as the only available future for Bhiwandi. At the same time, the geographies of logistics are also diffuse, ambiguous and continually shifting material sites that enable speedier and more efficient mobility of goods. Informational peripheries emerge from these gaps and overlaps between the mobilities and stickiness of logistics infrastructures. Informational peripheries emerge from the intersections between metropolitan peripheries where the underbelly of logistics finds itself, as well as the virtual peripheries where the underbelly of a digital age finds its only available future in an extractive economy of land, labour and information.

Methodology

This chapter is based on a two-year research project entitled 'Digitising the Periphery', whose larger aim was to investigate how peripheral municipalities implement national digitisation initiatives within systems of local governance, the struggles faced by marginal citizens in accessing public services in the peripheries, and the pathways they use to overcome these struggles and build resilience. The research team involved the author as principal investigator located in a British university, and a co-investigator and postdoctoral fellow based at the Tata Institute of Social Sciences, Mumbai. The project aimed to use methods such as surveys, semi-structured interviews and focus group meetings, and finally produce a toolkit for digital democracy and sustainable urbanisation.

Although initially warehouses were outside the scope of the research, since our case study was Bhiwandi (and the warehouses are located outside Bhiwandi's municipal boundaries), it soon became clear that if we were to understand the impacts of digitising the periphery, we needed to pay closer attention to the rapid transformations in the peripheries of Bhiwandi. It seemed that, particularly amongst the younger sections of Bhiwandi, employment in the warehousing sector was seen as the future. So, while the power loom industry was declining, younger people were turning their backs on this traditional sector and reaching out towards global logistical networks through the informational space afforded by digital infrastructures and technologies. As our interviews snowballed, we were led towards informants who had more to do with the warehousing sector than the power looms.

Perhaps this was to do with the gatekeepers who were approached. We started with a local vocational college, whose principal was a close friend of my co-investigator. The principal introduced us to a number of important stakeholders in Bhiwandi. Since Bhiwandi is a smaller urban centre with a Muslim-majority population, the most important stakeholders in the city are well networked with each other. Our respective positionalities – mine as an Indian and a professor at a British university, and my co-investigator from a minority community and a senior academic in a prominent Indian university, played a significant role in gaining access to, and the trust of, most of the participants. As the project started in 2021 during the COVID-19 lockdown, our initial interviews and some focus group meetings were conducted via Zoom. Once the lockdown was lifted and the research team could travel to Bhiwandi, we conducted an in-person focus group at the vocational college. There we were introduced to Saif, a young Muslim man who had become a successful broker in the warehousing sector. Saif spoke eloquently and had in-depth knowledge of the warehousing sector; he became one of our key informants. He took us on several visits to different types of warehouses and set up meetings with warehouse developers, managers and brokers, as well as locally elected officials. Later, Saif became a key participant in our stakeholder workshops.

All our participants were residents of Bhiwandi, and all the participants in the warehousing sector were men. We were unable to interview any warehouse workers because of the demanding nature of their work as well as the surveillance and control they faced, which increased their insecurities about answering questions related to their work.

Bhiwandi's peripheral futures

Figure 6.4 sets out the key moments in the history of Bhiwandi. Its origins as a municipality are tied to the rise of the British Empire and its subsequent transformation into a migrant town from the early nineteenth century. In 1857, after the Sepoy Mutiny[1] was violently suppressed by the British East India Company, thousands of Muslim weavers from North India travelled south along the Mumbai–Agra Highway to Bhiwandi. They brought with them their skills and their tools – the power looms, which established Bhiwandi and the region around it as one of the most prosperous textile manufacturing regions in the country. As the population rapidly increased, Bhiwandi was granted municipal corporation status in 1864. As Bhiwandi prospered over the next few decades, the nearby town of Nizampur was incorporated within its boundaries, and in 1918 it was named the Bhiwandi-Nizampur Municipal Corporation.

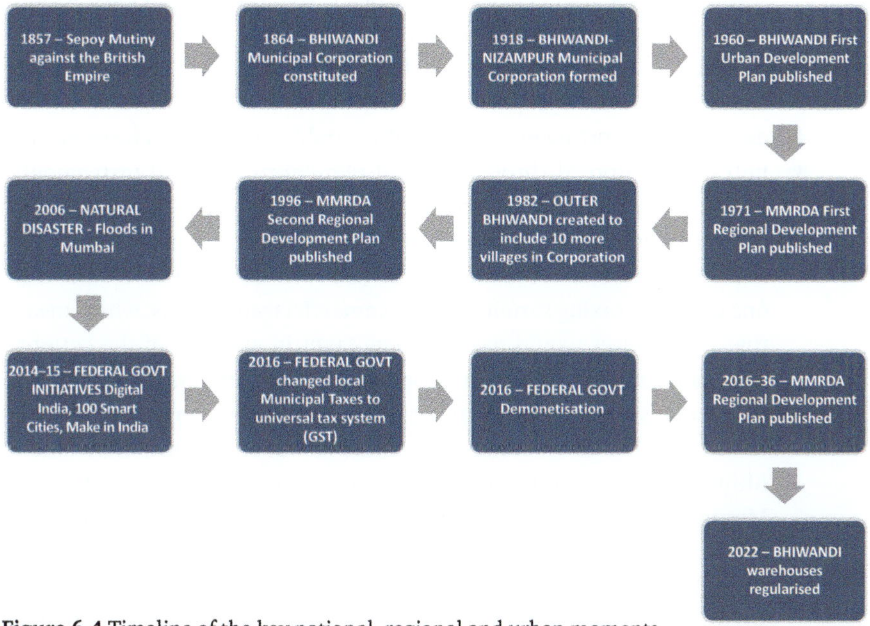

Figure 6.4 Timeline of the key national, regional and urban moments for Bhiwandi. Source: created by the author.

However, no comprehensive effort was made to plan the city until 1960, when its first Development Plan was created. The municipal boundary was further extended in 1982 to include 10 nearby villages, which are now collectively known as Outer Bhiwandi. While the power looms thrived and attracted migrant workers from across the country, the power loom sector itself remained largely unregulated and informal, characterised by cash-based transactions and employing thousands of circular migrant workers from across India. This 'informal' and unregulated nature of the power loom industry stemmed largely from its home-based model – families ran these businesses across generations, living on the upper floors of their homes while their power looms were on the ground floor. Today, Bhiwandi is home to approximately 650,000 power looms, accounting for roughly 33 per cent of the national total (Basu 2020). Despite its prosperity and significance as a national textile hub, it is this informal status which led to Bhiwandi being bypassed in successive Regional Development Plans produced by the Mumbai Metropolitan Regional Development Authority (MMRDA) in 1971 and 1996.

Around the early 1990s, warehouses began to relocate from Mumbai as commercial rents increased in the metropolis and Bhiwandi became a cheaper location to store goods, with its easy access to airports, seaports and the Mumbai–Agra Highway. In 2006, when Mumbai was hit with severe floods and many goods were destroyed, there was a mass exodus of warehouses from Mumbai to Bhiwandi. Around this time, highway networks were also expanded in the region, including the widening of the Mumbai–Nashik Expressway which passes through Bhiwandi. After a decade of supporting warehouses, Bhiwandi's strength as a warehousing hub was reinforced through several regulatory financial instruments. In 2016 the national government removed *octroi* – local sales taxes differentially levied by each regional state for industries and businesses – and installed General Standard Taxes (GST) across India. There was now one universal taxing system for all commercial transactions, which meant that warehouses would pay the same taxes in any part of the country. Thus, Bhiwandi became even more attractive for warehousing as national financial instruments pronounced its locational advantages.

Logistics looks for fungible land and labour to establish its stickiness. Bhiwandi's fungibility stemmed from its location in the Mumbai Metropolitan Region (MMR) with ready access to expressways, ports, and a rising middle-class urban population who were ready consumers of its goods. But its key fungibility was the availability of vast amounts of sizeable agricultural land parcels and a young unemployed population base who were reluctant to seek their future in the power loom sector.

Over the last few decades there has been a revolution of sorts in the spread and development of digital infrastructure in the MMR. With rising demand for more and better connectivity as India's economic capital, the Mumbai region now has some of the most advanced information and communication technologies in India. In recent years, with the launch of the national Digital India Initiative in 2015, the MMR has seen the setting up of large data centres and Business Processing Outsourcing operations in the peripheries of Mumbai and in the satellite township of Navi Mumbai. Digital infrastructure – such as fibreoptic cables, real-time communication, surveillance, barcoding, automations, heat management and robotics on par with international standards – which forms the foundations of advanced warehousing facilities, is therefore readily available in and around Bhiwandi. Alongside this, the entry of Middle Eastern steel companies such as Jameel Steel, which introduced mega-scale prefabricated steel construction, led to the transformation of the warehouse buildings from a modest 20,000 square feet to up to 600,000 square feet. Finally, the COVID-19 pandemic and the lockdown exacerbated a demand for online shopping and e-commerce that led to the mushrooming of warehouses since 2020.

The warehousing industry has transformed Bhiwandi into a 'supermarket for e-commerce players' (Borah 2022) and is now seen as Bhiwandi's future as well as the future of the MMR. While its older power loom industry has fallen into rapid decline, warehousing is deemed to be the industry of the future. The most recent 2016–36 Regional Development Plan from MMRDA mentions investment in warehouses as a future strategy, and indeed most of these warehouses were regularised in 2022.

Settler colonial logistics

On the day of our first workshop at a community college in Bhiwandi, several participants suggested that we meet Saif. We were unsure how significant Saif was for our project and what, if any, information we should expect from him. When Saif arrived at the workshop, albeit a little later than the others, he immediately began sharing in-depth knowledge about the history and evolution of warehousing around Bhiwandi. A broker in the warehousing industry, Saif was the protégé of an influential engineer-turned-developer from Bhiwandi. He spoke with pride about his mentor, who had been constructing godowns in Dubai, Mumbai and various airports since the 2000s, securing lucrative deals by undercutting competitors. His mentor then set his sights on Bhiwandi's old godowns, which he purchased, demolished and rebuilt with larger floor spaces. Currently, Saif's mentor owns around 700,000 square feet of godowns.

Warehouses in Bhiwandi are classified under the industrial sector, which enables ease of economic transactions and the generation of loans to the sector. Three regulatory mechanisms play a key role in extracting value from land and driving the production of warehouses. First is the agreement with the investor, who invests in both the land and the warehouse. Second is the agreement with the maintenance company, which supplies the physical and digital infrastructure and maintains them. Third is the agreement with the farmer, to whom these companies pay a nominal rent per square foot of warehouse space. When warehouses began relocating from Mumbai to Bhiwandi, agricultural land was the only possible fungible land that was available in large parcels. Initially, farmers were reluctant to sell their land, but the introduction of the *vadai* system – whereby farmers receive a portion of the rent generated by the warehouse, ensuring a continuous income – led to large-scale agricultural transformations. This arrangement sparked greater interest among large agricultural landholders to lease their land for warehouse development.

These three regulatory instruments enable various local and global companies to territorialise the 'logistics turn' in the region. Saif helped companies find these deals, acting as a middleman or broker in various transactions – negotiating with farmers, developers and contractors, as well as the companies that ultimately rent out the warehousing spaces. Saif's networks extended from the farming community to local politicians. He noted that companies from two hundred countries have warehouses in Bhiwandi, occupying 210 million square feet of warehouse area. This is not completely new: warehouses have been relocating to Bhiwandi since the 1990s. But since the early 2000s Bhiwandi has become a warehousing hub for Pan-Asia businesses because the rental costs are high in Mumbai ports and therefore it is cheaper to store goods in Bhiwandi. Saif noted that Bhiwandi's success relies on four Ds – Demand, Delivery, DHL (delivery company) and DB Schenker – the last being a German global logistics company which provides land transport, air freight and ocean freight to ship goods and supplies across the world.

Through these four Ds, three types of warehouses have emerged in Bhiwandi. First are the express warehouses, which operate 24/7 and typically store perishable goods related to groceries or e-commerce platforms like Big Basket, Kmart, Amazon and Flipkart. In this model, the goods come in and out mostly from the same region, and they are reliant on speedy connections. Second are the logistics warehouses, which occupy the largest warehousing spaces in the world, often covering around 700,000 square feet, and therefore also demand enormous amounts of power. Third are the custom-bonded warehouses, which are primarily focused on brand alteration and repricing and are therefore deterritorialised from the economic and regulatory constraints of the region. These warehouses store goods and brands that are not produced or sold in India, and therefore no customs and excise duties are payable in India. For example, products may be imported from China, re-labelled within the warehouse, and then exported to markets in the UK and Europe. In this model, Bhiwandi functions merely as a temporary holding place, facilitating the swift circulation of goods elsewhere.

These various types of warehouses embody what Porter and Yiftachel (2019) call the 'paradoxical' nature of settler colonialism, whereby warehouses occupy ancestral land held for generations in farming families, and in doing so also render the agricultural way of life in the region profoundly incompatible with the future of the region as a warehousing hub. This is a form of settler colonialism that lands with the technologies of temporality and yet settles as the future. However, this is not just about landing and circulating. Logistics as settler colonialism requires the input of several global and local actors in the process. As Cuppini and Frapporti note:

> Logistics delivers humans, animals, energy, earthly materials to an end, to a point, the point of production. But this includes, crucially, the point of production of the settler, the production of the entrepreneur, the banker, the slave trader, and the investor. (Cuppini and Frapporti 2018, 99–100)

The extractive value of land for warehouses in Bhiwandi is generated through deeply entrenched, intimate networks between businesses and their investors. Land itself and its territorialisation as warehouses are tied up with the privileges of landed caste and ethnic communities. Local communities engage in settler colonialism through explicit networks of wealth accumulation along all stages of the logistics economy. The land usually belongs to the Marathis, who are the ancestral landed castes and belong to the farming community. They receive a 15–20 per cent share in the rental value of warehouses as *vadai*. Local Marwari and Gujarati communities, who are from the merchant castes, do not hold land but invest in and own the buildings. The investors are the Gujarati Mahajans and the Kutchi Patels, who are ethnic and caste communities. But there are also semi-investors in the warehousing sector, mainly comprising logistics companies (such as DB Schenker and DHL). These companies maintain a limited market, investing mainly in renting space in the warehouses which they manage for e-commerce companies such as Amazon. Finally, there are the brokers – young men like Saif: entrepreneurs who facilitate the buying, selling and leasing of land and warehouses in the region.

It is possible to maintain these networks and the extractive value of land as these are all embedded in caste and religious affiliations that have developed in Bhiwandi over many generations. These affiliations have extended to the formal and informal communication systems between businesses, brokers and their investors. During the COVID-19 pandemic and lockdown, when investors were unable to meet face-to-face, the Community Tree app became a popular platform for communication via smartphone. Community Tree connected different castes and ethnic communities and has continued to be popular beyond the lockdowns.

Here, then, we see the blurring of the lines between peripheralisation, territorialisation and logistics compounded by the injection of informational spaces within local and global networks. Here information is not just the data on technologies, land and labour that may be available for logistics to take hold in Bhiwandi, but also the meaningfulness and value of information that circulates and spirals upwards – the type of information that is not readily available to global companies, but, when embedded within local networks, becomes a valuable asset for the expansion of global logistics.

Fungible labour

One scorching hot Ramadan afternoon, Ajmal and I were cooling off in his office. Ajmal is an entrepreneur who runs an 'educational institution', which in reality is more like a vocational college. He had been a small-time businessman until the COVID-19 pandemic forced him to close his business. Nearing bankruptcy, Ajmal received a lifeline when e-commerce companies began to flourish. Now, his primary livelihood comes from receiving commissions by matching floating labour in Bhiwandi with e-commerce companies, and he has recently employed seven staff just to facilitate this. Ajmal openly admitted that the system is highly exploitative, especially since warehouse workers are not allowed to unionise. He compared it to the gig sector, where employees can be hired and fired instantly. The desirable age for warehouse workers is between 18 and 38 – the younger the better. Engineers are always employed as data operators;

the less educated are employed in loading and unloading work. Workers need to have a very short, half-page CV. When a warehouse hires them, the words 'on trial' are stamped on the CV, and most positions last no longer than three months. Ajmal conceded that these practices are illegal under national employment laws, but the warehousing sector exploits a regulatory loophole. He is now organising a committee to challenge these practices. He said:

> Companies like migrant labour more because they are just working. The labourers from adjacent villages do not work for more than two to three months and they get tired and earn enough for pocket money and leave the job. But the migrant labour from Uttar Pradesh, Bihar or West Bengal migrate to earn. They are not distracted and only focus on work. They are here to earn, so they are preferred.

As we were talking, Ajmal's friend joined us. His name was Ali. He was another broker, like Ajmal, and administers over 250 WhatsApp lists for hiring warehouse labourers. He said each of his lists has over four hundred subscribers. Ali connects migrant workers with jobs. He claimed that he does not take a commission from workers, only from warehouses. But there is a lot of risk to this because if the worker does not stick with a job for at least two months, Ali does not receive his commission.

There is a whole ecosystem, then, of middlemen or employment brokers from the local Bhiwandi community who circulate fungible labour through the warehouses. Here 'fungible labour' refers to a particular embodied quality – migrant worker, non-unionised, fit and able to do manual work – making them interchangeable for the purposes of warehouse productivity. There are three types of labourers: those involved in unloading and loading, floor managers, and those overseeing placing and sorting. Additionally, some warehouses employ sweatshop labourers who manufacture products that are part of the logistics supply chain – car covers, toothbrushes, packing boxes – you name it, they have it. The demand for these jobs, particularly since the COVID-19 pandemic, further emphasises the interchangeable nature of the workforce. These workers exist in an informational periphery, with minimal access to information or rights, while information about them is obscured within the hundreds of WhatsApp lists administered by local brokers.

One locally elected official shared with us that migrant labourers in Bhiwandi struggle like 'slaves'. Although informality and precarity are also features of migrant labour in the power loom sector, this sector was, however, seen as more craft-based with training and skills development on the job. In the warehousing sector, labour is dispensable and interchangeable, and the e-commerce companies are complicit in creating precarity.

> A lot of these logistical parks have provided employment to the youth and given jobs but labour laws are not being implemented. But 95 per cent of the people are paid less than minimum wages but I don't want to touch that subject now because if I do that it will look anti-industry and people will not come to Bhiwandi. So I am letting them do that, settle down. Some companies have their own ethics but by the time it travels from the USA, to Europe, to India, it changes completely and becomes Indianised. It is a huge concern area. But it is true that labour exploitation is huge and it's an East India company kind of attitude by these multinationals. (B002)

This quote indicates how workers are kept in the informational periphery as fungible labour on fungible territory. The warehousing sector keeps both land and labour in an abstract future, portrayed as necessary for the future of the region – so much so that it is difficult for even elected officials to challenge this powerful narrative. The nature of employment in warehouses is abstracted to serve economic narratives of growth, and any challenge to this is seen as anti-developmentalist. A logistical approach is also seen in this continuous supply of labour, where a floating population of migrants is one of the many 'things' that flow in and out of the region to serve the needs of both global and local capital. This fungibility of labour requires an abstraction in the framings of supply, demand, flow and speed that are the hallmark of a logistical turn to the future of the region.

Territory, land and people in the margins of a digital age

In this chapter I have argued that urban studies scholarship needs to (re)position itself by paying attention to a logistical turn in the periphery. For this, we need to start rethinking urban studies from the periphery and then rethinking the periphery itself in a digital age. Peripheries are not just about informality or suburban sprawl; rather, they are made through uneven and unequal access to networked connectivity, as concentrated nodes of informational space emerge within longstanding structural disadvantages. As I have shown, logistics, digital infrastructures and a changing political economy of technology are remaking the periphery in the Global South. As the digital revolution takes root there, the periphery in this context reflects a 'spatial fix' of a global logistics network that is facilitated by informational access. Privileged information rests as much on a political technology of fungible land that makes it the ground zero of logistical territory as it does on a ready supply of fungible labour that is required to constantly move goods, on demand and in real time. But the periphery is more than this. It is re-emerging as a crucial site of regional futures, increasingly integrated into global networks of territory, logistics and people.

I have also argued that a settler colonial logistics in the periphery is as much entrenched within local networks as it is an external force of global capital. Settler coloniality works simultaneously through the regulatory practices and the intimate locational advantages of caste and ethnic privilege that enable warehouses to remain in the region as a permanent temporality. Settler colonial logistics requires fungible land and labour, and in Bhiwandi, the ready availability of both on account of vast swathes of agricultural land and migrant workers make it the hub of a 'settled' logistics. Settler colonial logistics also works through the squeezing out of the traditional power loom economy and the shifting of investment and interest towards warehousing as the future of the region.

This is what I have labelled 'informational periphery' (Datta 2024) – the paradoxical nature of informational flows, networks and stickiness in the urban and metropolitan peripheries of the Global South. The informational periphery is located both materially and epistemologically in the intersections of multiple imaginaries of the urban and digital, speed and decay, settling and moving. The informational periphery is a paradox. The informational periphery is a particular product of a logistics turn in urbanisation that focuses on speedy, on-demand, real-time information for some while withholding information from others. On the one hand, the periphery is radically shifting the territorial focus of metropolitan regions towards enabling a logistics model, whereby 'logistification' (Cavalier 2016) concentrates in declining industrial centres (such as Bhiwandi) to reinforce local privileges of caste, religion and ethnicity. On the other hand, it is produced by redlining the direction and flows of informational spaces. Digital transformations in the periphery deploy extractive technologies over people through the sedimentation of developmental delays, data deficits and information gaps. This has led to the production of fungible bodies and spaces which reflect what Simone has noted as the 'surrounds' – 'a space of insufficiency and incompletion' (Simone 2010, 40). The informational periphery, then, is built upon a global informational economy where Castells has argued 'power of flows takes precedence over the flows of power' (Castells 1996, 469).

Acknowledgements

An earlier version of this chapter was published as A. Datta, The informational periphery: Territory, logistics and people in the margins of a digital age (2024). *Asian Geographer* 41(2): 125–42. https://doi.org/10.1080/10225706.2023.2253233.

The material in this chapter was funded by the British Academy's Tackling Global Challenges research programme for the project 'Digitising the Periphery: Co-producing a toolkit for digital democracy and inclusive urbanisation' (ref: TGC\200118). I am grateful for the collaboration with Abdul Shaban and the contributions of Sheema Fatima, a postdoctoral research fellow at the Tata Institute of Social Sciences. I also thank the participants in Bhiwandi who patiently answered our questions and gave their support to this project.

Note

1 The Sepoy Mutiny is considered the first war of independence in India. It was led by the soldiers from the Indian subcontinent fighting in the British East India Company's army. Their aim was to reinstate the last Mughal emperor as the ruler of the Indian subcontinent. The uprising was violently crushed by the British East India Company; the Mughal emperor was sent to die in exile in Burma, and India was thereafter declared a colony under the British Crown.

References

Basu, B. 2020. Crisis in Bhiwandi powerloom sector. *Indian Textile Journal*, 19 June. Accessed February 2025. https://indiantextilejournal.com/crisis-in-bhiwandi-powerloom-sector.

Borah, Upamanyu. 2022. Bhiwandi: India's centralised stocking and distribution point (blog). Cargo Connect, 18 February. Accessed February 2025. http://www.cargoconnect.co.in/hub/bhiwandi-indias-centralised-stocking-and-distribution-point.

Castells, Manuel. 1996. *The Network Society: The information age*. Oxford: Blackwell.

Cavalier, Jesse. 2016. *The Rule of Logistics: Walmart and the architecture of fulfillment*. Minneapolis, MN: University of Minnesota Press.

Chua, Charmaine S. 2017. Logistical violence, logistical vulnerabilities. *Historical Materialism* 25(4): 167–82. https://doi.org/10.1163/1569206X-12341544.

Chua, Charmaine S. 2021. Lineages of infrastructural power: The logistical nightmare of Los Angeles. In *Between Catastrophe and Revolution: Essays in honor of Mike Davis*, edited by Daniel Bertrand Monk and Michael Sorkin, 249–68. New York: OR Books.

Chua, Charmaine S., Martin Danyluk, Deborah Cowen and Laleh Khalili. 2018. Introduction: Turbulent circulation: Building a critical engagement with logistics. *Environment and Planning D: Society and space* 36(4): 617–29. https://doi.org/10.1177/0263775818783101.

Cidell, Julie. 2011. Distribution centers among the rooftops: The global logistics network meets the suburban spatial imaginary: Distribution centers in suburban Chicago. *International Journal of Urban and Regional Research* 35(4): 832–51. https://doi.org/10.1111/j.1468-2427.2010.00973.x.

Cowen, Deborah. 2014. *The Deadly Life of Logistics*. Minneapolis, MN: University of Minnesota Press.

Cuppini, Niccolò and Mattia Frapporti. 2018. Logistics genealogies: A dialogue with Stefano Harney. *SocialText* 36(3): 95–110. https://doi.org/10.1215/01642472-6917802.

Datta, Ayona. 2023. The digitalising state: Governing digitalisation-as-urbanisation in the global south. *Progress in Human Geography* 47(1): 141–59. https://doi.org/10.1177/03091325221141798.

Datta, Ayona. 2024. The informational periphery: Territory, logistics and people in the margins of a digital age. *Asian Geographer* 41(2): 125–42. https://doi.org/10.1080/10225706.2023.2253233.

Du, Yue. 2019. Urbanizing the periphery: Infrastructure funding and local growth coalition in China's peasant relocation programs. *Urban Geography* 40(9): 1,231–50. https://doi.org/10.1080/02723638.2019.1568811.

Easterling, Keller. 2016. *Extrastatecraft: The power of infrastructure space*. London: Verso.

Furlong, Kathryn. 2020. Geographies of infrastructure II: Concrete, cloud and layered (in)visibilities. *Progress in Human Geography* 45(1): 190–8. https://doi.org/10.1177/0309132520923098.

Jordan, Tim. 2015. *Information Politics*. London: Pluto Press.

Kanai, J. Miguel and Seth Schindler. 2018. Peri-urban promises of connectivity: Linking project-led polycentrism to the infrastructure scramble. *Environment and Planning A: Economy and space* 51(2): 302–22. https://doi.org/10.1177/0308518X18763370.

Keith, Michael and Andreza Aruska de Souza Santos. 2021. *African Cities and Collaborative Futures: Urban platforms and metropolitan logistics*. Manchester: Manchester University Press.

Lyster, Clare. 2016. *Learning from Logistics: How networks change our cities*. Basel: Birkhauser.

Mutter, Samuel. 2023. Distribution, dis-sumption and dis-appointment: The negative geographies of city logistics. *Progress in Human Geography* 47(1): 160–77. https://doi.org/10.1177/03091325221132563.

Porter, Libby and Oren Yiftachel. 2019. Urbanizing settler-colonial studies: Introduction to the special issue. *Settler Colonial Studies* 9(2): 177–86. https://doi.org/10.1080/2201473X.2017.1409394.

Shaban, Abdul, Ayona Datta and Sheema Fatima. 2023. Overlapping marginalities: Digital education and Muslim female students in Bhiwandi. *Economic & Political Weekly* 58(29). https://www.epw.in/journal/2023/29/special-articles/overlapping-marginalities.html.

Simone, AbdouMaliq. 2010. *City Life from Jakarta to Dakar: Movements at the crossroads*. Abingdon: Routledge.

Simone, AbdouMaliq. 2022. *The Surrounds: Urban life within and beyond capture*. Durham, NC: Duke University Press.

Yiftachel, Oren. 2009. Critical theory and 'gray space': Mobilization of the colonized. *City* 13(2–3): 246–63. https://doi.org/10.1080/13604810902982227.

Zeiderman, Austin. 2021. In the wake of logistics: Situated afterlives of race and labour on the Magdalena river. *Environment and Planning D: Society and space* 39(3): 441–58. https://doi.org/10.1177/0263775820970945.

Zins, Chaim. 2007. Conceptual approaches for defining data, information, and knowledge. *Journal of the American Society for Information Science and Technology* 58(4): 479–93. https://doi.org/10.1002/asi.20508.

7
Informational peripheries as ecological peripheries: vignettes from Guiyang

Calvin King Lam Chung, Jingya Dai and Jiang Xu

Not all spaces in and around cities have benefited equally from the arrival of the information age. Moving beyond 'a focus on centers of technological design and development' (Chan 2013, 8), research on information and digital geographies (Ash et al. 2018; Graham et al. 2014) has paid increasing attention to the production of informational peripheries: territories that are marginalised and dispossessed by and for particular patterns of informational flows for the benefit of some (Datta 2023). While much has been written about the political ambitions and social inequalities underpinning these informational peripheries (Banerji 2024; Datta, Chapter 6 this volume), few have examined how the peripheralisation of such territories is also an *ecological* process, involving changes in the flows and stocks of materials and organisms in the biophysical environment (Walker 2005).

This chapter will shed light on the ecological dimension of informational peripheries by drawing attention to how local biophysical conditions are reshaped by the multitude of information flowing through them. As we will elaborate, informational peripheries can simultaneously be 'ecological peripheries' (Ross 2020), whose biophysical environment suffers due to their peripherality in existing regimes of information governance and development. We identify two types of *informational-ecological peripheries*, each reflecting a particular form of entanglement between information and the environment. The first type concerns territories lying on the geographical periphery of the reach of informational infrastructure for environmental governance.

These territories, lacking environmental monitoring infrastructure, are at greater risk of environmental mismanagement and exploitation. Lifting them out of environmental 'information poverty' (Graham et al. 2014) with the aid of digital technologies has emerged as a priority for many governments facing growing pressure to protect the environment. The second type concerns territories that, although not bypassed by informational flows, remain peripheral to the informational economy they support. These regions host critical infrastructure that keeps information as well as the capital and goods they mobilise flowing. Their peripherality is defined not only by their distance from those 'flagship urban spaces' of smart development (Caprotti 2019), but also by the high ecological costs they incur, often invisible to the beneficiaries of the informational economy.

To illustrate these informational-ecological peripheries, this chapter turns to recent developments in Guiyang, the capital city of the southwest Chinese province of Guizhou. Once an economically peripheral city on the national scale, Guiyang has gained widespread fame for its rapid transformation into China's 'big data valley', as well as being a national environmental exemplar (Chung and Xu 2021). Since the 2010s, the local government of Guiyang has touted the importance of its city being green and digital at the same time, as if they are two sides of the same coin. On the one hand, the Guiyang government harnesses digital technologies to strengthen its oversight of the city's peripheries of environmental information – that is, areas poorly covered by the existing environmental monitoring system. On the other hand, the government has also turned its city into an informational periphery featuring clusters of data centres that, given their significant resource demands, may not be as green as they have been touted.

Our subsequent analysis of the making and unmaking of informational-ecological peripheries in Guiyang is based on a synthesis of our research work conducted from 2016 to 2024 in three research projects covering the city and other parts of the Guizhou province. These projects focus on environmental discourses in urban planning, local implementation of data-driven environmental governance, and the growth of the big data sector, respectively. We reviewed a range of documents, including speeches given by local government officials, policy papers and plans by the national, Guizhou and Guiyang governments, and publications on Guiyang's recent economic and environmental transformations. We also interviewed over a hundred people involved in or familiar with Guiyang's environmental governance and digital development, including local and provincial officials, urban planners, researchers affiliated with government units or universities, and entrepreneurs in the big data sector. Although the informational periphery was not a guiding concept for any of our research projects, its potential as an analytical lens to make sense of Guiyang's recent development has surfaced within the repeated references in the documents we reviewed and the words of our interviewees to Guiyang, and sometimes Guizhou as a whole, as a 'backward' and 'marginal' territory that can be upgraded through information.

Informational peripheries, the state and the environment

Informational peripheries of governance and development

Debates about the stark core–periphery patterns produced by the information revolution have a history as long as the information revolution itself (Hepworth and Robins 1988). Rather than enjoying 'all information in all places at all times' (Godfrey 1979) as the technological optimist expected, we have learned from the expanding literature on information geographies the highly uneven and unequal ways in which information technology has transformed spaces politically, economically and socially (Ash et al. 2018; Graham et al. 2014). To map and explain this socio-spatial disparity of information, a growing rank of scholars has mobilised the notion of the informational periphery, which, in its various guises, denotes those territories that have been marginalised by and for informational flows (Datta 2023; Graham et al. 2014; Mol 2009).

Existing studies have alluded to two types of informational peripheries. The first type concerns the spatial outcomes of 'governing through information' (Soma et al. 2016), or the use of information as a means of governance. Some of these territories have not been digitised into information and are therefore less knowable and governable, while others have been deliberately bypassed and misrepresented to render them manipulable (Datta 2023). They are produced by a medley of constraints: economic ones, which hinder the construction and maintenance of information infrastructure; political ones, which distort the collection, processing and dissemination of information; and cultural ones, which discredit the value of information for governance (Mol 2009). The recent eruption of initiatives to augment cities and regions with digital technologies reflects a technocratic utopianism wherein territories can be governed more efficiently and effectively through an expanding array of technologies to measure, monitor and intervene in socio-spatial processes (Kitchin 2014), although scepticism abounds regarding the transformative changes these initiatives may bring – an empirical question still warranting more investigation in different parts of the world.

The second type of informational peripheries are upshots of the territorialisation of what Castells (2010) refers to as 'informational capitalism', marked by the growing centrality of information networks in how economic activities are conducted. This renewed form of capitalism produces peripheral territories in the sense that they are transformed for the 'cores of informational affluence' (Shaw and Graham 2017, 911) – companies and cities which accumulate through their monopolies in capturing and circulating information – to enable both the flows of information and the flows of goods and capital such information seeks to manoeuvre. In Chapter 6 of this volume, Datta offers a vivid illustration of this type of informational peripheries with her study of Bhiwandi, a town located on the fringes of Mumbai, where mega-scale warehouses have mushroomed to serve the logistics needs of the growing business of hundreds of e-commerce companies based elsewhere. While the first type of informational periphery is information-poor, Bhiwandi is arguably more embedded in the global flows of information that move a growing diversity and quantity of goods through its warehouses. Nonetheless, as with those non-digitised and disinformed peripheries, it benefits little from the information revolution and is vulnerable to exploitation because of its lack of control over the information that shapes it.

The digitalising state: national and local

Studies have highlighted the pivotal role of the state in the production of new information geographies (Kitchin 2023), including both of the foregoing types of informational peripheries. In the realm of governance, many states are said to be digitalising, drawing upon an expanding repertoire of digital technologies to expand their reach and deepen their influence (Datta 2023; Schou and Hjelholt 2019). This process of state-led digitalisation has found its most intense expression in those informational peripheries marked by a 'relative scarcity of maps, survey statistics and census data in the pre-digital era' (Datta 2023, 151), as the state unmakes their informational peripherality by inscribing a variety of digital infrastructures in order to enhance the visibility, intelligibility and hence governability of their people and places. At the same time, to ride the wave of informational capitalism, many states are also eagerly digitalising their economy by rolling out strategic plans and investment aids to promote the development of digital industries (Foster and Azmeh

2020). Apart from social media and platform-based services (for example, ride-hailing and food delivery services) that are closest to the end users, digital industries also include a suite of businesses which provide the technological and infrastructural foundations of a digital economy, such as telecommunications, high-performance computing and artificial intelligence firms (Zook 2020). In this new economy, data and the information derived from them are valued as capital to be accumulated (Sadowski 2019).

Whether the emphasis is on digitalising governance or the economy, it is not just the national governments that are at play. Local governments have often steered the implementation of such digitalisation agendas as they compete to brand and develop their jurisdictions as 'smart cities' (Karvonen et al. 2018). For many cities, to be smart is to apply digital technologies wide and far, leaving no corners of a city as informational peripheries of governance. Notions such as 'sensor cities' (D'Amico et al. 2020) and 'dashboard urbanism' (Odendaal 2019) testify to the growing centrality of technology-aided collection, processing and analysis of data in a ubiquitous, constant and exhaustive manner, in the hope that the information so derived about urban dynamics can inform better urban policy making. However, for some cities, to be smart means to achieve economic growth through digital and informational means. As part of their technical fix to urban entrepreneurialism, local states can be seen proactively improving both hard and soft infrastructure of their cities to compete for investments from digital giants and anchor start-ups (Clark 2020; McGuirk et al. 2021). The physical and functional integration of digital technologies into urban processes has become a crucial means for cities to reimagine themselves as favourable testbeds of new technologies to attract an even wider range of high-tech economic sectors (Barns and Pollio 2018). Of course, not all cities can capture the core and high value-added activities of the digital value chain, such as those of digital production and application; for many, their digital industry is limited to hosting data centres which serve the data storage and computing needs of other cities with a booming digital economy.

The ecological dimension of informational peripheries

So far, research on informational peripheries has reflected a strong interest in the socio-spatial inequalities engendered or reinforced by the uneven territorialisation of information (Banerji 2024; Datta, Chapter 6 this volume). However, as a growing body of political ecology literature reminds us, the space-making impacts of informational flows are no less ecological.

Political ecologists have long emphasised the need to 're-materialise' (Berkhout and Hertin 2004) digital technologies, whose ostensible virtuality – promoted, for instance, by the language of the 'cloud' (Atkins 2021) – has prevented many people from seeing them as a potential cause of environmental problems. These scholars have revealed a multitude of flows of matter and energy that are set in motion by and for the digital production and consumption of information, such as the pollution caused by lead, mercury and other toxic chemicals released by disposed electronics (Gabrys 2011), and the scramble for control over supplies of minerals that are crucial to manufacturing computers (Ensmenger 2018). More recently, the explosive development of data technologies has garnered an outpouring of research interests in information as 'powerful materialisers' (Marvin 1997, 47) which, in its different forms and use, can make a huge difference to the environment (Nost and Goldstein 2022) – for example, how advances in telecommunications have generated new demands for travel and hence new emissions (Marvin 1997). In this sense, the territorialisation of information is as much about reshaping the socio-economic landscape as it is about reshaping the environment.

As a corollary, we propose that informational peripheries can simultaneously be ecological peripheries. Here we are indebted to Ross, who approached the process of peripheralisation as 'an interaction between human and non-human forces' (Ross 2020, 467). Citing examples of colonial plantations, he argued that parts of Asia, Africa and the Americas have been turned into economic peripheries as they functioned as what he called ecological peripheries, supplying cheap natural resources to and absorbing ecological costs of production for the Global North. As for informational peripheries, we argue that their ecological dimension can be explored by examining two different forms of information–environment entanglement.

The first entanglement finds expression in the biophysical implications of approaches to governing the environment with and through information. The information revolution has generated a new set of informational dynamics in the field of environmental governance, notably an upsurge in optimism about the future of environmental protection through more environmental monitoring activities at multiple spatial scales. These activities aim to track environmental changes, such as shifts in biodiversity and air quality, and audit the compliance of a multitude of state and non-state actors with their environmental commitments, such as conservation efforts and emissions control (Mol 2009). With this emphasis on environmental information, many parts of the world can be classified as informational peripheries at risk of environmental mismanagement and exploitation because of their historical lack of informational infrastructure to monitor the environment. However, given the convergence of state digitalisation agendas with the goals and targets of sustainable development, recent years have seen widespread national and local state efforts to eliminate existing environmental informational peripheries by means of deploying an expanding assemblage of environmental sensors, cloud-based platforms and diagnostic algorithms to collect, process and analyse environmental information, all meant to render the metabolism of cities and regions more visible and amenable to intervention (Nost and Goldstein 2022; Viitanen and Kingston 2014).

Another entanglement takes the form of the environmental impacts of the very act of digital production and consumption of information within the framework of informational capitalism. Although digital industries are often seen as part of a clean and green economy (Gibbs and Krueger 2007), recent studies have highlighted their 'extractive, resource-intensive' nature (Pickren 2018, 237; Lucivero 2020). The mushrooming of warehouses for e-commerce firms on agricultural land (Datta, Chapter 6 this volume) is an example of the biophysical transformations that are part and parcel of a developing digital economy. For many scholars, though, those informational peripheries emerge as the result of the large-scale development of data centres, which serve as both the infrastructural backbone and competitive businesses of the informational age. Brenner and Schmid (2015, 167) would have referred to these peripheries as 'operational landscapes': territories that are geographically removed from urban centres and opened up to meet the socio-metabolic and – we shall add – informational imperatives associated with urban growth. In these territories, the construction of data centres involves dispossessing and transforming vast stretches of land (Faxon and Kintzi 2022), while their operation requires a stable and substantial round-the-clock supply of electricity to power their computing and storage systems and to cool these systems down (Brevini 2020; Brodie 2023). As data centres continue to expand with the digital economy, preventing them from gobbling up more electricity and generating more carbon emissions has become a pressing environmental issue (Jones 2018).

These cases reveal that informational peripheries can also be ecological peripheries, whose environment is either peripheral to governance or exploited as a periphery in the service of a centre's development. As the state engages with information in different ways, it makes and unmakes different informational peripheries entangled with different environmental problems. When the state invests in informational infrastructure to enhance governance, such as environmental sensors and platforms, it may eliminate informational peripheries as hotspots of informational poverty to allow their environment to be better monitored for protection. However, when the state embraces information for capital accumulation, it engenders informational peripheries where the environment is exploited for the digital storage and processing of information. In other words, informational peripheries are ecological peripheries where the environment is worse off by having either too little information to govern it or too much information circulating through it.

Contextual roots of Guiyang's big data turn

China offers an instructive window into the complex entanglements of information and the environment, as its central government throws them into stark relief through its concomitant pursuit of environmental sustainability and digitalisation. In recent decades, the Chinese state has increasingly focused on environmental issues. Since the 1990s, the central government has introduced a series of new laws and policies aimed at curbing environmental degradation. The introduction of the vision to develop China into an 'ecological civilisation' by then-President Hu Jintao in 2007 – since emphasised more forcefully by President Xi Jinping – marks the latest effort by China's leaders to make environmental protection a goal that should be contemplated in all aspects of governance, including economic and urban governance (Kostka and Zhang 2018). To realise this vision, the central government is an adamant adherent of the informational governance paradigm, believing that increased access to environmental information is a prerequisite for better environmental governance (Kostka et al. 2020). In 2015 it mandated the construction of a nationwide environmental monitoring network (State Council 2015), which involves establishing more monitoring stations on the ground, increasing the use of satellites for remote monitoring, and centralising the management of data collected by these means in the hands of the national environmental agencies. Doing so will allow the central government to keep a close eye on local environmental changes and appraise local officials' performance based on accurate environmental information. If this network is successfully established, it will eliminate any informational peripheries in China's environmental governance.

This emphasis on increasing the use of digital technologies for state environmental regulation coincides with the Chinese state's ambition to upgrade China into a digital nation – one that mobilises the power of digital technologies to accelerate innovation-driven economic growth. The high political priority of this high-tech turn is marked by ambitious national science and technology plans such as 'Digital China' and 'Made in China 2025', aimed at promoting indigenous innovation. Specifically, the Chinese state has identified information and communication technologies as a key emerging industry that aligns with its industrial upgrading agenda, alongside other sectors such as biotech, renewable energy and new energy cars. To support these industries, the state has provided substantial support through tax incentives and special funding (Xu et al. 2022). In February 2023, while this chapter was being drafted, the Chinese State Council announced that local government officials would be assessed based on the digitalisation achievements within their jurisdictions. As a result, the digital economy has gained traction. According to a report from the Chinese State Council, China's digital economy generated 45.5 trillion RMB in 2021 – over a third of that year's national GDP (Xinhua 2022). Many of China's economically peripheral cities have become eager to align their governance goals with this digital focus, with a view to being favoured by the central government in its policy making and resource allocation, in addition to attracting related private investment (Xu and Chung 2024) – a trend that contrasts with findings elsewhere which more frequently attribute changes in urban peripheries to neoliberal forces (Gururani and Kennedy 2021). Specifically, local governments have rushed to embrace smart urbanism in digitalising their governance and economy (Zhou et al. 2023). Just as the Chinese state hopes to turn China into a core country in the informational age, these local governments are keen to help their cities gain a seat in – to borrow Shaw and Graham's words – the country's 'cores of informational affluence' (Shaw and Graham 2017, 911).

The central government's dual emphasis on making the country both green and digital has placed significant pressure on the local government of Guiyang, a city once labelled as poor, backward and environmentally vulnerable. Given its inland and mountainous location, Guiyang has long been an economic backwater. When China's economic reforms began in 1978, the city was not blessed with the same favourable policies in taxation, trade and investment that its coastal counterparts received as testbeds of market-oriented approaches to development (Chung and Xu 2021). During the 1980s and 1990s, Guiyang's economy remained as resource-dependent as it had been before the reform, dominated by the coal, phosphate and aluminium industries. This was despite the fact that Guiyang is an ecologically fragile city: more than 80 per cent of its aquifers consist of carbonate rocks, which have high permeability, leaving local groundwater vulnerable to contamination from industrial development (Lang et al. 2006). The city's extensive coal burning also led to severe air pollution. However, the turn of the millennium saw a sharp change in Guiyang's development trajectory, as the local state steered the city's future to align with the central government's vision. In response to an environmentalising central government, the Guiyang government launched large-scale urban greening and clean-up initiatives, positioning itself ahead of other local states by declaring ecological civilisation construction as its overarching governance goal. Echoing the central government's digitalisation plan, Guiyang has led the country in embedding the collection and analysis of 'big data' – the catch-all term for large and complex datasets that can be mined for immense amounts of information (Xu et al. 2023) – in and for every aspect of life. These endeavours have attracted the central government's attention, earning Guiyang important national recognition. In 2012 it was designated as China's first pilot city for the new national goal of building an ecological civilisation. Then, in 2015, it became one of the first officially supported cities to develop agglomerations of big data firms under the central government's national big data action plan.

Against this policy backdrop, the Guiyang government has endeavoured to eliminate the city's last vestiges of peripherality and keep it ahead of the curve of China's green and digital transition. In the next two sections, we examine two specific moves made by the local state at the intersection of information and environment. First, it has sought to draw on big data technologies to eradicate Guiyang's informational peripheries, as ecological spaces that are poorly covered by the city's environmental monitoring system. Second, it has committed to anchoring big data firms in Guiyang, notably those running data centres, as an apparently green strategy for the city's growth – a move that has peripheralised Gui'an as a zone of ecological exploitation for the digital economy's informational infrastructure.

Eradicating Guiyang's peripheries of environmental information

Contrasting with its positive image today, the Guiyang government was noted – like those of many other less developed cities in China – for its lack of technical capacity to carry out essential environmental monitoring and regulation tasks. Guiyang's environmental officials were sarcastically portrayed in the mass media as 'monitoring with eyes, noses, and ears'

(Huang 2016). To make matters worse, local corporations in Guizhou tended to be perfunctory in addressing pollution emissions, a problem that remained the case even recently. For example, in the environmental inspections conducted by the central government in 2022, it was found that some firms were continuing to discharge untreated sewage into nearby streams, while others had failed to undertake ecological restoration on mines that were no longer in operation (Guizhou Government 2022). Therefore, since the 2010s, the Guiyang government has proactively harnessed an expanding repertoire of digital technologies (for example, geographic information systems (GIS), networked sensors, cloud platforms and smart wearables) to promote environmental protection and conservation, in the hope that smart technologies will equip environmental officials with 'brilliant ears' (*shunfeng er*) and 'far sight' (*qianli yan*) (Huang 2016).

In Guiyang, the district of Wudang has piloted this digitalisation of environmental regulation. Previously, district-level government officials could only gauge general trends in the environmental quality of their jurisdictions based on data from a handful of environmental monitoring stations, not to mention tracking down the sources of any pollution. Beyond the immediate areas covered by these stations, many parts of Wudang were effectively informational peripheries for the district's environmental managers. However, in 2016, the district government constructed an 'environmental data grid system' (Guizhou Investment Promotion Bureau 2016; Hsu et al. 2020). This grid system divides Wudang's 686 kilometre-square territory into multiple grids, each equipped with sensors that constantly collect ambient environmental data (such as air quality, water quality and noise levels), which are then uploaded to an ecological big data system for real-time analysis. A total of 266 monitoring sites with environmental sensors have been set up across the district. In the event of a pollution episode, this grid system enables rapid and precise detection of pollution sources through its big data algorithms, which can estimate which of the key pollution sources (such as construction sites, restaurants or factories) in the grid concerned is most likely to be responsible for the crisis.

The use of digital technologies has also transformed environmental governance, promptly redressing environmental problems and achieving effective coordination among environmental officials. For instance, since October 2019 the Guiyang government has been operating a big data system for river management so that local water regulation can go paperless. With the aid of a River Chief Cloud app, 'river chiefs' – officials in charge of the water quality of a given river (Wang and Chen

2020) – can browse real-time water quality data on their rivers using their mobile phones. If an environmental official spots floating sewage in a river, they can report the issue through the app (Interview, April 2023). The corresponding river chief can then, again through the app, dispatch their subordinates to address the issue promptly. As one river chief commented, 'every task [of river regulation] can be completed through the [River Chief] Cloud' (*Guiyang Daily* 2019; Hsu et al. 2020). Moreover, the Guiyang government has made the River Chief Cloud app publicly accessible, in the hope that local people will serve as 'citizen scientists' and report issues relating to bodies of water. In this regard, the use of digital technologies can facilitate dialogue between governments and the general public on environmental governance, and promote the transparency and accountability of the local government's work on environmental protection. Thanks to these new digital means, Guiyang has cleared 29 black and smelly water bodies in its city, and the water quality of some branches has increased from 'below Grade V' (the lowest tier of China's six-tier water quality assessment system, with Grade I being the best quality) to Grade III (suitable for drinking purposes) (National Development and Reform Commission 2021).

The digitalisation of Guiyang's environmental governance is not limited to pollution reduction. With over half of its territory covered by forest (*China News* 2021), the Guiyang government has sought to take advantage of digital innovations to enhance the visibility of biophysical processes within its forests, which have long formed part of the city's informational periphery in environmental governance due to their distribution across rugged terrains. A report by China National Radio (2022) affords us a glimpse into the city's forest management systems: at one of Guiyang's forest fire command centres, a large screen displays multiple live feeds of forests from various angles. Behind the screen, a set of algorithms detects blazes from these live streams, triggering fire alerts to forestry officials in the control centre. In the event of a fire, officials can deploy drones to determine the exact location of the blazes for prompt firefighting action. Guiyang has over 50 such fire command centres, each monitoring around 20 square kilometres of forest and detecting fires with pinpoint accuracy, down to the nearest metre (*Tianyan Xinwen* 2020). Now that the Guiyang government can manage its peripheral forests remotely, the annual extent of the city's forests affected by fire has been significantly reduced.

Developing Guiyang into an ecological periphery of big data

Guiyang's green framing of the big data sector

In the economic realm, the Guiyang government has upheld the development of the big data sector as a win-win approach to the city's growth imperatives and environmental aspirations in the informational age. Although the term is used somewhat loosely in popular contexts in China, the big data sector is generally understood as an emerging economic sector consisting of firms engaging in the acquisition, analysis, curation, storage and use of large, complex datasets (Curry 2016; Xu et al. 2023). Guiyang's big data sector development can be traced back to 2013, when Chen Gang became the city's party secretary. Before this, Chen was a member of the party standing committee of Beijing, where he oversaw the development of Zhongguancun Technology Park, China's first national high-tech development zone. His governance expertise, coupled with his vision for urban futures driven by digital innovation, enabled him to lay the groundwork for developing Guiyang's big data sector.

Chen's vision of orchestrating a big data boom in Guiyang can be understood as a response to the city's natural conditions. On the one hand, both Guiyang and its provincial neighbours face the challenge of developing a clean industrial structure due to local environmental constraints. This concern was highlighted at the 2012 Guiyang Eco Forum, an annual conference dedicated to exploring ways of constructing an ecological civilisation in Guiyang and beyond. Wu Hequan, an academic from the Chinese Academy of Engineering and a participant in the 2012 forum, recalled: '[W]e reached a preliminary consensus, that is, Guizhou's ecology is fragile and cannot withstand destruction. The province cannot continue developing intensive industries. What can we develop instead? The emerging big data sector might be a promising direction' (*Technology Daily* 2021). Chen elaborated on this point during China's annual parliamentary meetings in 2014, where he explained why the big data sector is fundamental to the development of an environmentally fragile province:

> On the surface, Guiyang appears to have pristine waters and verdant mountains [which gives people an impression of a good environment]. But in fact, Guiyang features karst topography, which brings a big challenge that the local environmental carrying capacity

> is very weak. Therefore, to develop the economy and safeguard the environment at the same time, the answer is straightforward – we must adopt an innovation-driven approach. (Ifeng 2014)

On the other hand, Chen recognised Guiyang's environmental advantages for hosting big data firms and has often boasted about these qualities in public interviews. Situated at elevations ranging from 875 to 1,655 metres above sea level, Guiyang has a year-round cool climate, with an average annual temperature of around 15 degrees Celsius (Lang et al. 2006; see Figure 7.1). This is important for the siting of big data firms: a relatively low temperature helps reduce the energy required to cool servers, thereby lowering energy costs (Nost and Goldstein 2022). Meanwhile, Guiyang's rugged karst landscape, while hindering large-scale agricultural and urban development, has unexpectedly turned the city into an optimal location for data centres, which can take advantage of the natural cooling effects of strong winds flowing through intermountain passes, further reducing the need for air conditioning. Therefore, Guiyang offers a greener choice for big data firms compared to other cities with higher temperatures. Given that energy bills constitute a large proportion of their operating costs (Lucivero 2020), big data firms are attracted by these potential cost savings.

Another key reason for Guiyang to embrace big data development was that it would be a pivotal strategy for the city to achieve economic revitalisation while simultaneously meeting local environmental objectives. A quote from Chen's interview with *People's Daily* in 2015 illustrates his views:

> The ecological environment in Guiyang is good, so we should not take the old trajectory of 'pollution first, treatment second'. Neither should we take the path of progressive industrial upgrading and expansion. From the perspective of development, Guiyang needs to understand the laws of development of high-tech and modern manufacturing industries in order to understand [its potential]. (*People's Daily* 2015)

For Chen, 'high-tech' industries, to which the big data sector belongs, are presumably 'green' industries, or else he would not have contrasted a development trajectory led by such industries with the dirty old one he mentions earlier. He reiterated the environmental motivations behind his decision that Guiyang should venture into the big data sector in a 2018 interview with *Tencent News*:

Figure 7.1 A banner in Guiyang erected during the city's Big Data Expo 2023. The banner reads 'China Data Valley – Cool Guiyang'. Photograph: Calvin Chung.

> Everyone knows that China's development has been rapid in recent years, but it has also led to a common social issue: while the economy has grown, the environment has been damaged ... Therefore, the central government has paid particular attention to which development trajectory Guiyang will take. That trajectory should be one driven by new technologies – the development cannot be slow, and the environment cannot be damaged. It must be led by innovation. In this case, big data is our important lever. (*Tencent News* 2018)

The materiality of Guiyang's big data sector

Chen's discursive greening of the big data sector seemed to be well received by the Guizhou government. In 2016, it issued a green industry directory specifying several key industries to be promoted as green businesses in the province, with the big data sector being one of them (State Council 2016). However, if we turn our attention from Guiyang's environmental advantages to the big data sector, and to the ecological footprint of this sector on the city, we may draw a rather different conclusion.

One thing to note about the structure of Guiyang's big data sector is the predominance of data centres. All of China's big three telecoms carriers – China Mobile, China Unicom and China Telecom – built their data centres in Guiyang. Other technological giants have followed suit, including Alibaba, Apple, Baidu, Huawei and Tencent. However, due to a lack of digital talent, few of these companies have established a significant presence in Guiyang for other stages of the big data value chain, particularly data analysis research and application; these stages are still carried out in the traditional and predictable hubs of economic growth, such as Beijing, Shanghai, Shenzhen, Hangzhou and Chongqing (Interview, August 2023). It is also the firms in these economically core cities that Guiyang's data centres serve. In this regard, we might argue that Guiyang is trapped in the informational periphery, marked by ecological extraction for the core.

First of all, data centres consume land. When we speak of data centres that have flocked to Guiyang, most of them are found in Gui'an, a 1,795 kilometre-square special policy district established with notable autonomy in 2014 by merging the southwestern part of Guiyang with the northeastern corner of the neighbouring city of Anshun. Since 2018, Gui'an has been under the administrative oversight of the Guiyang government. Gui'an has been designated to house these data centres and their associated businesses because it is the province's most extensive area of flat land, which reduces the geoengineering challenges of constructing these facilities. However, this has meant that the farmland that once dominated the locality has been erased. According to a Chinese state media report (news.cctv.com 2021), Gui'an is now home to 12 super-large data centres occupying 7 square kilometres. This footprint is set to increase as the central government has designated Gui'an as one of the 10 national data centre clusters in its plan for a national computing network (*China Daily* 2022), which will reinforce the role of Gui'an as a periphery providing data storage and computing power for businesses in the east. In addition, as a local official revealed, compared to other local jurisdictions in Guizhou, Gui'an has a much smaller proportion of its territory bounded by the ecological redline, behind which industrial activities are prohibited for conservation reasons (Interview, December 2017). This preferential treatment may be read as a sign that even the provincial government – which determines each jurisdiction's redlined areas – has permitted the big data sector to achieve runaway growth under the green smokescreen that the Guiyang government has presented.

Secondly, the Guiyang government's efforts to stimulate the development of the big data sector through favourable policies are not free from environmental concerns. Electricity pricing is a case in point. Tech firms establishing their data centres in Guiyang enjoy a favourable tariff of 0.35 yuan/kWh thanks to subsidies from the local government (Guizhou Development and Reform Commission 2016; 2020). In comparison, the average tariff for data centres in China is 0.87 yuan/kWh, while data centres in eastern China consume electricity at 1.5 yuan/kWh, four times that of their counterparts in Guiyang (Sina News 2018). If this pricing structure raises worries among environmentalists about energy inefficiency incentivised by cheap electricity, they might be further disturbed by the possibility that the massive amount of energy that supports the operation of the big data sector does not come from green sources. While there is no publicly available information on Guiyang's energy mix for electricity generation, it is known that, as of 2021, as much as 45.8 per cent of Guizhou's installed capacity for electricity generation was still associated with coal-fired power plants (Power China 2021). Meanwhile, the share of installed capacity associated with solar and wind power was 22.7 per cent (Power China 2021). Perhaps the big data sector is environmentally more benign when compared to the traditional industries that Chen Gang discredited as too dirty for Guiyang, but its energy footprint seems to undermine Chen's claim about the supposedly green trajectory safeguarded by big data. So far, the disjuncture between the green forays of the big data sector rehearsed by the Guiyang government and the potential environmental adversities entailed by its development has been downplayed by optimism that data centres will improve energy efficiency. Nonetheless, it is the adoption of a greener energy mix that will make a real difference to their environmental impacts.

Entangled peripheries

The recent development of Guiyang is a story of peripheries in many senses. To its local government, Guiyang was an economic periphery that needed to be put on the national map to earn more support from the central government and attract more investment. To achieve this goal, the Guiyang government has sought not only to advance sustainability and digitalisation as separate realms but also to combine their agendas. The ensuing territorialisation of information of different kinds and for different reasons has meant that informational peripheries are being made and unmade at the same time in Guiyang, albeit in different forms.

With respect to how information has been digitally harnessed in Guiyang, we distinguish two types of informational peripheries. The first type is the result of the uneven territorialisation of information as a governance apparatus; to be peripheral is to be bypassed by informational infrastructure and kept in informational poverty. To unmake such informational peripheries in its city, such as poorly monitored airsheds, rivers and forests, the Guiyang government has endeavoured to govern through digitally augmented information collection and dissemination. The other type is the result of the spatial division of labour of an informational economy; to be peripheral is to be subservient to attempts in other territories to accumulate through information. Guiyang is developing into this type of informational periphery as it becomes a hub of data centres, storing and processing data from more developed cities.

The case of Guiyang also enlightens us with regard to the entanglements of informational peripheries and the environment: both of the foregoing types of informational peripheries can concomitantly be ecological peripheries at risk of exploitation of the biophysical environment. In Guiyang, the local state-led digitalisation, or big data, agenda has paradoxically improved some aspects of the local environment while increasing the peripheralisation of others. On the one hand, the local government's widespread use of digital technologies in environmental governance has lifted many parts of Guiyang's ecosystem out of the ecological periphery by first lifting them out of the periphery of environmental information. On the other hand, information is environmentally detrimental to Guiyang when it is stored and processed by the city's growing fleet of data centres, which rely on considerable amounts of land and electricity. By embracing an informationally peripheral future as a national data centre hub, Guiyang risks entrenching itself in the ecological periphery that serves a materially extractive digital economy.

Although we have analysed the two types of informational peripheries in Guiyang separately, we do not preclude the possibility that they might overlap. For example, excluding a site from environmental monitoring – thus creating an informational periphery of governance – might be a way for the state to prepare it as a tabula rasa for developing a periphery of informational capitalism populated by data centres or e-commerce warehouses. As digitalisation continues apace in various directions, one can only imagine that future studies will reveal an even broader range of informational peripheries and their entanglements with the environment than those we have learned from Guiyang.

Acknowledgement

The research reported in this chapter was supported by the Research Grants Council of Hong Kong (ref: 24610420, 14616421 and 14637516).

References

Ash, J., Kitchin, R. and Leszczynski, A. (eds). 2018. *Digital Geographies*. Thousand Oaks, CA: Sage Publications.

Atkins, E. 2021. Tracing the 'cloud': Emergent political geographies of global data centres. *Political Geography* 86. https://doi.org/10.1016/j.polgeo.2020.102306.

Banerji, S. 2024. Māṇḍavlī: negotiating with digital governance in Mumbai. *Urban Geography* 45(3): 460–83. https://doi.org/10.1080/02723638.2023.2200309.

Barns, S. and Pollio, A. 2018. Parramatta smart city and the quest to build Australia's next great city. In *Inside Smart Cities: Place, politics and urban innovation*, edited by A. Karvonen, F. Cugurullo and F. Caprotti, 197–210. Abingdon: Routledge.

Berkhout, F. and Hertin, J. 2004. De-materialising and re-materialising: Digital technologies and the environment. *Futures* 36(8): 903–20. https://doi.org/10.1016/j.futures.2004.01.003.

Brenner, N. and Schmid, C. 2015. Towards a new epistemology of the urban? *City* 19(2–3): 151–82. https://doi.org/10.1080/13604813.2015.1014712.

Brevini, B. 2020. Black boxes, not green: Mythologizing artificial intelligence and omitting the environment. *Big Data & Society* 7(2). https://doi.org/10.1177/2053951720935141.

Brodie, P. 2023. Data infrastructure studies on an unequal planet. *Big Data & Society* 10(1). https://doi.org/10.1177/20539517231182402.

Caprotti, F. 2019. Spaces of visibility in the smart city: Flagship urban spaces and the smart urban imaginary. *Urban Studies* 56(12): 2465–79. https://doi.org/10.1177/0042098018798597.

Castells, M. 2010. *End of Millennium*. 2nd edn. Oxford: Blackwell.

Chan, A. S. 2013. *Networking Peripheries: Technological futures and the myth of digital universalism*. Cambridge, MA: MIT Press.

China Daily. 2022. National computing network to synergize east and west. 25 February. Accessed February 2025. http://global.chinadaily.com.cn/a/202202/25/WS62181cb9a310cdd39bc88c93.html.

China National Radio (Yangguang Wang). 2022. Guiyang: Cheng 'yun' ershang, caliang 'dashuju' mingpian [Guiyang: Rise with cloud, renowned for big data]. 3 November. Accessed February 2025. https://gz.cnr.cn/jiaodiantu/20221103/t20221103_526049791.shtml.

China News. 2021. Zhongguo shouge guojia senlin chengshi guiyang: Senlin fugailv da 55% [Guiyang, first national forestry city in China: Forest coverage rate reached over 55 per cent]. 7 January. Accessed February 2025. https://www.guiyang.gov.cn/ztzl/rdzt/ztzlwmcsrmcwmcgrrx/ztzlwmcsrmcwmcgrrxcwdt/202103/t20210324_67306622.html.

Chung, C. K. L. and Xu, J. 2021. Scalar politics of urban sustainability: Governing the Chinese city in the era of ecological civilisation. *Transactions of the Institute of British Geographers* 46(3): 689–703. https://doi.org/10.1111/tran.12436.

Clark, J. 2020. *Uneven Innovation: The work of smart cities*. New York: Columbia University Press.

Curry, E. 2016. The big data value chain: Definitions, concepts, and theoretical approaches. In *New Horizons for a Data-Driven Economy: A roadmap for usage and exploitation of big data in Europe*, edited by J. M. Cavanillas, E. Curry and W. Wahlster, 29–37. Dordrecht: Springer.

D'Amico, G., L'Abbate, P., Liao, W., Yigitcanlar, T. and Ioppolo, G. 2020. Understanding sensor cities: Insights from technology giant company driven smart urbanism practices. *Sensors* 20(16). https://doi.org/10.3390/s20164391.

Datta, A. 2023. The digitalising state: Governing digitalisation-as-urbanisation in the global south. *Progress in Human Geography* 47(1): 141–59. https://doi.org/10.1177/03091325221141798.

Ensmenger, N. 2018. The environmental history of computing. *Technology and Culture* 59(4): S7–S33. https://doi.org/10.1353/tech.2018.0148.

Faxon, H. O. and Kintzi, K. 2022. Critical geographies of smart development. *Transactions of the Institute of British Geographers* 47(4): 898–911. https://doi.org/10.1111/tran.12560.

Foster, C. and Azmeh, S. 2020. Latecomer economies and national digital policy: An industrial policy perspective. *Journal of Development Studies* 56(7): 1247–62. https://doi.org/10.1080/00220388.2019.1677886.

Gabrys, J. 2011. *Digital Rubbish: A natural history of electronics*. Ann Arbor, MI: University of Michigan Press.

Gibbs, D. and Krueger, R. 2007. Containing the contradcitions of rapid development? New economy spaces and sustainable urban development. In *The Sustainable Development Paradox: Urban political economy in the United States and Europe*, edited by R. Krueger and D. Gibbs, 95–122. New York: Guilford Press.

Godfrey, D. 1979. All information in all places at all times. In *Gutenberg Two*, edited by D. Godfrey and D. Parkhill. Toronto: Press Porcépic.

Graham, M., Hogan, B., Straumann, R. K. and Medhat, A. 2014. Uneven geographies of user-generated information: Patterns of increasing informational poverty. *Annals of the Association of American Geographers* 104(4): 746–64. https://doi.org/10.1080/00045608.2014.910087.

Guiyang Daily (Guiyang Ribao). 2019. Dashuju funeng shengtai baohu de wudang shijian [Big data fuel ecological conservation practices in Wudang, Guiyang]. 31 May. Accessed February 2025. https://www.sohu.com/a/317846631_120025455.

Guizhou Development and Reform Commission. 2016. Guanyu jiangdi dagongye qiye yongdian chengben cujin zhuanxing shengji de shishi fang'an [Implementation plan on reducing electricity costs of large industrial enterprises and promoting transformation and upgrading]. Accessed February 2025. https://www.ndrc.gov.cn/xwdt/ztzl/gdqjcbzc/guizhou/201806/P020191205483716120243.pdf.

Guizhou Development and Reform Commission. 2020. Sheng dashuju fazhan guanli ju guanyu shengzhengxie shier jie sanci huiyi di 1189 hao ti'an de huifu [Reply to Proposal No. 1189 on the third session of the 12th Guizhou Province Political Consultative Conference].

Guizhou Government. 2022. Guizhou sheng guanche luoshi di'er lun zhongyang shengtai huanjing baohu ducha baogao zhenggai fang'an [Guizhou Province implements the rectification plan for the second round of central ecological and environmental protection inspection report]. 16 September. Accessed February 2025. https://www.mee.gov.cn/ywgz/zysthjbhdc/dczg/202209/t20220917_994228.shtml.

Guizhou Investment Promotion Bureau. 2016. Guiyang shouge shengtai huanjing dashuju shidian xiangmu zai wudang kaishi yunxing [First ecological environmental big data pilot project starts in Wudang district]. 20 July. Accessed February 2025. https://www.rockontrol.com/mtbd/3127.jhtml.

Gururani, S. and Kennedy, L. 2021. Engaging the urban from the periphery: The co-production of space, politics and subjectivities in India's urban peripheries. *South Asia Multidisciplinary Academic Journal* 26. https://doi.org/10.4000/samaj.7131.

Hepworth, M. and Robins, K. 1988. Whose information society? A view from the periphery. *Media, Culture & Society* 10(3): 323–43. https://doi.org/10.1177/016344388010003005.

Hsu, A., Yeo, Z. Y. and Weinfurter, A. 2020. Emerging digital environmental governance in China: The case of black and smelly waters in China. *Journal of Environmental Planning and Management* 63(1): 14–31. https://doi.org/10.1080/09640568.2019.1661228.

Huang, Y. 2016. Guizhou huanbao kaiqi 'yunshang moshi' [Guizhou's environmental protection switches on 'cloud mode']. *China Environment News*, 20 December.

Ifeng (Fenghuang Wang). 2014. Chen Gang: Guizhou fazhan pingjing duo, dashuju keyi liantong sihai [Chen Gang: Guizhou faces many development bottlenecks, big data could connect the world]. Accessed February 2025. http://phtv.ifeng.com/program/wdsz/detail_2014_03/28/35250848_0.shtml.

Jones, N. 2018. How to stop data centres from gobbling up the world's electricity. *Nature* 561(7722): 163–6. https://doi.org/10.1038/d41586-018-06610-y.

Karvonen, A., Cugurullo, F. and Caprotti, F. (eds). 2018. *Inside Smart Cities: Place, politics and urban innovation*. Abingdon: Routledge.

Kitchin, R. 2014. The real-time city? Big data and smart urbanism. *GeoJournal* 79(1): 1–14. https://doi.org/10.1007/s10708-013-9516-8.

Kitchin, R. 2023. Urban data power: Capitalism, governance, ethics, and justice. In *Data Power in Action: Urban data politics in times of crisis*, edited by O. Söderström and A. Datta, 21–41. Bristol: Bristol University Press.

Kostka, G. and Zhang, C. 2018. Tightening the grip: Environmental governance under Xi Jinping. *Environmental Politics* 27(5): 769–81. https://doi.org/10.1080/09644016.2018.1491116.

Kostka, G., Zhang, X. and Shin, K. 2020. Information, technology, and digitalization in China's environmental governance. *Journal of Environmental Planning and Management* 63(1): 1–13. https://doi.org/10.1080/09640568.2019.1681386.

Lang, Y. C., Liu, C. Q., Zhao, Z. Q., Li, S. L. and Han, G. L. 2006. Geochemistry of surface and ground water in Guiyang, China: Water/rock interaction and pollution in a karst hydrological system. *Applied Geochemistry* 21(6): 887–903. https://doi.org/10.1016/j.apgeochem.2006.03.005.

Lucivero, F. 2020. Big data, big waste? A reflection on the environmental sustainability of big data initiatives. *Science and Engineering Ethics* 26(2): 1,009–30. https://doi.org/10.1007/s11948-019-00171-7.

Marvin, S. 1997. Environmental flows: Telecommunications and the dematerialisation of cities? *Futures* 29(1): 47–65. https://doi.org/10.1016/S0016-3287(96)00065-1.

McGuirk, P., Dowling, R. and Chatterjee, P. 2021. Municipal statecraft for the smart city: Retooling the smart entrepreneurial city? *Environment and Planning A: Economy and space* 53(7): 1730–48. https://doi.org/10.1177/0308518X211027905.

Mol, A. P. 2009. Environmental governance through information: China and Vietnam. *Singapore Journal of Tropical Geography* 30(1): 114–29. https://doi.org/10.1111/j.1467-9493.2008.00358.x.

National Development and Reform Commission. 2021. Tongchou tuijin, jingzhun zhiwu, nanminghe huanqing yumin – guiyang shi tuijin nanminghe zhili qude shixiao [Coordinated and targeted pollution control makes Nanming River clear again – Guiyang achieved results in Nanming River treatment]. Accessed February 2025. https://www.ndrc.gov.cn/fggz/hjyzy/sjyybh/202111/t20211105_1303096.html.

news.cctv.com. 2021. Dashuju fazhan kan Guizhou: Duoge qiye xiangji luohu, Gui'an dazao daxing shuju zhongxin juji qu [Seeing big data development through Guizhou: Multiple companies have successively settled in Gui'an, creating a large data centre cluster]. 26 May. Accessed February 2025. https://news.cctv.com/2021/05/26/ARTI6v1473ULXWbdcGOg7pU0210526.shtml.

Nost, E. and Goldstein, J. E. 2022. A political ecology of data. *Environment and Planning E: Nature and space* 5(1): 3–17. https://doi.org/10.1177/25148486211043503.

Odendaal, N. 2019. Appropriating 'big data': Exploring the emancipatory potential of the data strategies of civil society organizations in Cape Town, South Africa. In *Right to the Smart City*, edited by P. Cardullo, C. Di Feliciantonio and R. Kitchin, 165–76. Leeds: Emerald Publishing.

People's Daily. 2015. Guiyang chuangxin qudong zhilu: Cong 'zhongguancun yuansu' dao 'zhongchou jinrong' [Guiyang's innovation-driven path: From Zhongguancun elements to crowdfunding finance]. Accessed February 2025. https://money.people.com.cn/n/2015/1021/c42877-27723961.html.

Pickren, G. 2018. 'The global assemblage of digital flow': Critical data studies and the infrastructures of computing. *Progress in Human Geography* 42(2): 225–43. https://doi.org/10.1177/0309132516673241.

Power China. 2021. Yitu kandong guizhousheng nengyuan fazhan baogao 2021 [Guizhou Energy Development Report 2021]. Accessed February 2025. https://mp.weixin.qq.com/s/Lp1WIbTkvmsYxXUB-M7cxg.

Ross, C. 2020. Nature, labour, and the making of ecological peripheries. *International Review of Social History* 65(3): 467–79. https://doi.org/10.1017/S0020859020000565.

Sadowski, J. 2019. When data is capital: Datafication, accumulation, and extraction. *Big Data & Society* 6(1). https://doi.org/10.1177/2053951718820549.

Schou, J. and Hjelholt, M. 2019. Digital state spaces: State rescaling and advanced digitalization. *Territory, Politics, Governance* 7(4): 438–54. https://doi.org/10.1080/21622671.2018.1532809.

Shaw, J. and Graham, M. 2017. An informational right to the city? Code, content, control, and the urbanization of information. *Antipode* 49(4): 907–27. https://doi.org/10.1111/anti.12312.

Sina News 2018. Guizhou dashuju chanye tanmi: jigou xuqiu wangsheng, guipiao zu jueqi [Explore Guizhou's big data sector: Strong demands from institutions, big surge in Guizhou drifters]. 30 November. Accessed February 2025. http://k.sina.com.cn/article_1686546714_6486a91a02000j062.html?cre=tianyi&mod=pcpager_fintoutiao&loc=22&r=9&doct=0&rfunc=100&tj=none&tr=9.

Soma, K., Termeer, C. J. and Opdam, P. 2016. Informational governance: A systematic literature review of governance for sustainability in the information age. *Environmental Science & Policy* 56: 89–99. https://doi.org/10.1016/j.envsci.2015.11.006.

State Council. 2015. Shengtai huanjing jiance wangluo jianshe fangan [Construction plan of ecological environment monitoring network]. Accessed February 2025. http://www.gov.cn/zhengce/content/2015-08/12/content_10078.htm.

State Council. 2016. Guizhou sheng fabu dashengtai shida gongchengbao he lvse jingji "sixing" chanye fazhan yindao mulu [Guizhou government released 10 major big ecological projects and Industrial Development Guidance Directory for Four Main Green Industries]. Accessed February 2025. http://www.gov.cn/xinwen/2016-10/11/content_5117248.htm.

Technology Daily. 2021. Qinian qiande jueze gei zhege xibu shengfen dailai le shenme? [What has decision made seven years ago brought to this western province?]. 9 June. Accessed February 2025. http://www.stdaily.com/cehua/kbstjj/2021-06/11/content_1155578.shtml.

Tencent News. 2018. Guiyang shiwei shuji Chen Gang: Dashuju shidai de Guizhou [Guiyang party secretary Chen Gang: Guizhou in the era of big data]. Accessed February 2025. https://cloud.tencent.com.cn/developer/article/1131224.

Tianyan Xinwen. 2020. Guiyang: 'Dashuju+' kaiqi senlin fanghuo zhihui moshi [Guiyang: 'Big data+' activates a smart model for forest fire prevetion]. 31 March. Accessed February 2025. https://baijiahao.baidu.com/s?id=1662652179465961100.

Viitanen, J. and Kingston, R. 2014. Smart cities and green growth: Outsourcing democratic and environmental resilience to the global technology sector. *Environment and Planning A: Economy and space* 46(4): 803–19. https://doi.org/10.1068/a46242.

Walker, P. A. 2005. Political ecology: Where is the ecology? *Progress in Human Geography* 29(1): 73–82. https://doi.org/10.1191/0309132505ph530pr.

Wang, Y. and Chen, X. 2020. River chief system as a collaborative water governance approach in China. *International Journal of Water Resources Development* 36(4): 610–30. https://doi.org/10.1080/07900627.2019.1680351.

Xinhua. 2022. Shuzi jingji yi chengwei tuidong zhongguo jingji zengzhang zhu yinqing zhiyi [Digital economy becomes one of the key engines driving China's economic growth]. Accessed February 2025. http://www.gov.cn/xinwen/2022-11/07/content_5725115.htm.

Xu, J. and Chung, C. K. L. 2024. Of scale and discourse: China's ecological civilization and the struggle for development in Anshun. *Urban Geography* 45(6): 986–1005. https://doi.org/10.1080/02723638.2023.2243141.

Xu, J., Li, A., Chung, C. K. L. and Yue, Y. 2023. Mapping the unmapped: Investigating big data companies via online sources. *Professional Geographer* 75(5): 816–26. https://doi.org/10.1080/00330124.2023.2169175.

Xu, J., Zou, G. and Chung, C. K. L. 2022. (Re)shaping urban governance through state–business interaction in inland China's emerging industries. *The China Review* 22(4): 77–104.

Zhou, Y., Xiao, F. and Deng, W. 2023. Is smart city a slogan? Evidence from China. *Asian Geographer* 40(2): 185–202. https://doi.org/10.1080/10225706.2022.2052734.

Zook, M. 2020. Industries. In *Digital Geographies*, edited by J. Ash, R. Kitchin and A. Leszczynski, 188–99. Thousand Oaks, CA: Sage Publications.

8

Follow the data: computing the algorithmic periphery with credit scores and property values

Julien Migozzi

Digital technologies are transforming cities. Urban peripheries such as South African townships – the archetype of state-enforced segregation (Western 1981) – reflect the reconfiguration of urban economies and everyday lives in the era of digital capitalism: it takes a couple of clicks to notice the presence of these areas on property portals such as PrivateProperty or Property24. Every month, around 50 million users browse these online marketplaces, filtering properties by price, location and features. Networking buyers, properties and real-estate agents, these platforms integrate luxury neighbourhoods and impoverished townships into a single architecture, where abundant information comes in all data shapes: myriad numbers, rich textual descriptions, and piles of pictures and statistical widgets showing local market trends.

The technical possibility of seamlessly navigating this digital rendering of the post-apartheid city contrasts with the urban history of South Africa, where the housing market was harshly instrumentalised since the colonial era to separate and enforce racial segregation, generating long-lasting wealth inequalities. Under apartheid, Black South Africans were denied property rights in urban areas, and people of colour were forcefully relegated into urban peripheries; tasked with fostering White homeownership, building societies and banks mostly catered for the demographic minority while redlining townships. Nowadays racial discrimination is banned, and these same banks purchase marketing banners on property portals to advertise their three-minute online mortgage applications.

Like other sectors such as financial services and logistics, the housing market has been deeply reconfigured by the centrality of data and the rise of the platform economy in South Africa (Migozzi 2024a; Pollio and Cirolia 2022). Yet how this roll-out of digital technologies by both state and non-state actors affects the reconfiguration of urban spaces remains overlooked, despite the powerful role of housing markets in shaping urban forms and hierarchies. How do the conversion of land, housing and people into digital data, and the market practices associated with the information they convey, affect geographical inequalities and urban marginality? How are these informational flows, backed up by an avalanche of data yet inherently socially and spatially selective, producing informational peripheries through the networking of the material and digital worlds (Datta 2024)?

To answer these questions, this chapter adopts 'follow the data' as a method to investigate the digitisation and informational flows of the post-apartheid housing market, characterised by the large-scale circulation of consumer and property data. My contribution is twofold. First, I propose 'follow the data' as a method to examine the infrastructural anatomy, regimes of value creation and mechanisms of stratification that characterise urban geographies under digital capitalism (Birch 2023; Burrell and Fourcade 2021; Galster 2023; Sadowski 2020). Second, since informational flows play a key role in the production of informational peripheries (Datta 2024), I use follow the data to study the content, directionality and selectivity of these flows. Seeing and following data as 'digitally encoded information' (Ash et al. 2024) opens up empirical and theoretical ways to both disentangle the infrastructural plumbing and techno-political economy of informational flows and explain how they articulate inequalities of knowledge, wealth and power in the reconfiguration of urban spaces.

This chapter investigates the political economy and stratifying effects of informational flows that govern the post-apartheid housing market, following two data streams respectively fed by the credit reporting system and the digitisation of title deeds. Following these streams illuminates the daily functioning and calculative turn of the housing market, to reveal how homeseekers and homes become unequally seen and categorised by real-estate actors through the extraction, circulation and assetisation of data. The output of these informational flows – that is, credit scores and property values – shapes urban segregation and hierarchies: in the context of racialised indebtedness and extreme housing inequalities (James 2014; Migozzi 2020), these metrics are used on an everyday basis by real-estate agents and banks to sort out customers, assess mortgage applications and categorise neighbourhoods. Informational flows therefore enforce a new regime of visibility across the post-apartheid city: under the formal market's algorithmic gaze, people and homes need to be identified and 'available for measurement' (Fourcade and Healy 2017, 19). Yet these flows are the product of racialised dispossession and inequalities, they come with data gaps, and their output consequently perpetuate biases towards precariousness and informality. Poor and indebted citizens, along with properties located in townships and informal settlements, are cast in the algorithmic periphery – ranked at the bottom of data-driven, ordinal categories (Fourcade and Healy 2024) that either exclude from the housing market or include on unfavourable terms. While integrated into the market's stack through their conversion into digital data (Masiero 2023), people associated with urban marginality are reframed by informational flows as digital outcasts, categorised as 'incalculable', 'invisible' or 'disqualified'.

The first section presents 'follow the data' as a method: by defining a starting point, a directionality and a sectoral scope, it becomes possible to sequence and model data flows across infrastructures and organisations, and identify crossing points and practices of data assetisation. The second section examines the identifiers that support the circulation of 'indexical data' (Kitchin 2014) – data that enable identification and linkages of datasets, underlining the informational legacies of colonialism and apartheid (Migozzi 2024a). I then unpack the institutional anatomy of two data streams – one for people, one for properties. At key crossing points, consumer and deeds data enter proprietary grounds: harvested, stored, analysed and commercialised, data are transformed into income-bearing assets for credit bureaus and analytics companies. The fourth section explains how the ensuing calculation of credit scores and property values constitutes the output of the market's informational flows, used every day by real-estate actors to sort customers and make lending decisions, affecting access to housing and residential mobilities. The fifth section examines how people and houses associated with urban marginality are cast in the algorithmic periphery of low scores and values, categorised by informational flows as 'invisible', 'incalculable' or 'disqualified'. Bringing into the conversation the scholarship on redlining (Aalbers 2005; Taylor 2019), the sixth and final section suggests a refinement of the empirical and theoretical lenses of the concept of the informational periphery, which needs to go beyond its focus on logistics and its definition of infrastructural redlining as bypassing or exclusion: I discuss how foregrounding the role of identification systems and the process of unequal inclusion is essential to better capture the making of informational peripheries.

Follow the data: starting and crossing points, sectoral scope and directionality

Following data flows across space and time opens up analytical possibilities to understand how digital technologies reshape urban spaces and livelihoods. Despite a wealth of studies on networks, platforms and big data, few studies, especially in in geography, offer a methodology to analyse how data actually circulate and affect practices, despite early calls to centre the role of big data (Kitchin 2013) and the wider infrastructural turn of the discipline (Graham and Marvin 2001). This might come as a surprise given that following 'things', such as commodities or financial flows, forms an integral part of the methodological toolkit to unpack the functioning of markets and disentangle power structures in relations of productions and exchanges (Cook 2004; Christophers 2011; Taffel 2023).

Yet following data generated by and circulated through digital technologies comes with significant empirical and definitional challenges. Data can be defined as a 'manufactured material that intrinsically has value' (Rob 2021), whose collection and analysis further drive people, firms and governments into the data-driven world of digital capitalism (Fourcade and Healy 2024). As data stand at the crossroads of manufacturing, valuation and decision making, follow the data opens up many questions: how are data generated and networked among actors and across infrastructures? How are data turned into economic assets? And, with regard to informational peripheries, how are data interpreted as legible information to shape the practices and imaginaries through which geographical inequalities are recomposed? Crucially, it is not only about *how*, but also about *where* and *when*: since data are inherently spatial (Dalton et al. 2016), the method of following needs to pay particular attention to places and temporalities in order to locate 'on what terms people are enrolled in regimes of data generation and collection' (Dalton et al. 2016), how these regimes affect processes of value creation, and, in return, how they affect organisation models, political governance and everyday lives.

Yet the nature of data in the twenty-first century might defy the most seasoned tracker. The volume and velocity of the digital traces left by human activities (Salganik 2017) make it impossible for a human eye to embrace big data – we can only see bits and parcels of the networked manufacturing process: small samples, extracts and queries, aggregated outputs, snapshots of lines of code. Big data are also 'always-on' – in a permanent state of production and circulation; access is difficult due to proprietary, multi-sited and relational databases; black-boxed algorithms and changing engineering systems obfuscate the mechanisms of data creation – and the patterns they create (Salganik 2017). Since data are shapeshifting, found everywhere and always-on, one needs to adopt strategies to define the scope, directionality and perimeters of the investigation. In that regard, follow the data shares a methodological approach with 'follow the money', a common method in the discipline of geography. Since money too escapes clear-cut definitional boundaries, scholars use proxies to measure and map the volume and directionality of money flows, choosing a sectoral scope, and collecting digital traces of specific events: bond payments, mergers & acquisitions transactions, accounting lines etc. (Hughes-McLure 2022; Kass 2020; Wojcik et al. 2024).

Defining a starting point, a directionality and a sectoral scope opens up analytical potential to follow the data. Inspired by the scholarship on commodity chains, Akbari chooses the event of a personal video that failed to circulate during the 2009 uprising in Iran; following the traces left by the video, she reveals an 'intricate assemblage of organisations, policies, laws, code, software, people, and platforms' (Akbari 2020, 424). To unpack the political economies of internet infrastructures in Northern Europe, Flensburg and Lai follow data through the 'infrastructural layers of digital communication' (Flensburg and Lai 2023, 322), starting with 'the first stop' of the internet – local access networks such as optical fibre, mobile frequencies, satellites etc. Bates et al. introduce the concept of 'data journeys' (places of data production, processing, distribution and use) to map out the socio-material practices of UK weather data across sites and actors: starting from measuring instruments located in weather stations, they then trace how data moves across different 'sites of data practises' (Bates et al. 2016, 1–2). Common to the last two studies is the use of visuals to map and model data flows. What is, however, missing in all these attempts is using the follow-up process to examine a key dimension of digital capitalism: when and how data become an economic asset – a resource and a property which provide regular revenue streams (Birch et al. 2020).

Building upon these approaches, I choose a generic event, which is bounded geographically and temporally yet subject to many iterations: a bonded sale, wherein an ordinary buyer purchases a house with a mortgage. This type of transaction is the outcome of informational flows networking real-estate agents, banks and homebuyers, but is also the source of further data creation and circulation. Follow the data is therefore particularly suited to examining the digital assemblages that reshape real estate, an industry historically characterised by a strong appetite for data (Migozzi 2024b; Sawyer et al. 2014). It is also a timely method for identifying the crossing points of extraction and circulation, through which data become subject to enclosure and assetisation (Birch 2023): real estate has indeed been the subject of multiple digital experiments that have platformed housing and land markets, reshaping their socio-technical construction and political economy (Cowan 2021; Fields 2024; Nethercote 2023).

This method of follow the data builds upon a decade of in-depth fieldwork on the real-estate industry in Cape Town to map and model the data streams triggered by a bonded sale. I conducted 116 interviews with stakeholders all along the market's value chain (buyers, real-estate agents, developers, mortgage originators, banks, data brokers, credit bureaus, attorneys), and shadowed agents in a real-estate agency located in Khayelitsha, Cape Town's largest black township, to better understand the paper-based and digital practices involved at every step of the transaction process. Buying a house might represent a trivial event, yet each step of the process (finding a property, obtaining a mortgage, signing an offer to purchase and a deed of sale, registering the title deed etc.) rests upon a flow of data, fed by manual entries and automated queries.

Observing and inventorying the data produced, circulated and interpreted by market actors, I propose a visual model of the data flows that govern the South African housing market (Figure 8.1). Like any model, it answers a need for simplification and abstraction (Chorley and Haggett 1967). Using the case of a bonded sale, I model the functioning of the housing market in two main data streams: one streams circulates information on people, the other on properties. I then follow data in two directions: upstream, looking at the data flows whose output regulates and conditions the transaction process; and downstream, looking at the data generated by the finalisation of the transaction. Converting land, housing and people into digital data, these informational flows recompose urban inequalities.

Data identifiers: South Africa's informational legacies

To follow data, one needs to look at the identifiers that support the creation, circulation and querying of data during the sale process. Identifiers play a crucial role in informational flows: as 'labels of physical and digital objects and services', identifiers are 'part of how we make sense of the world and communicate' (Open Data Institute 2020). In South Africa, two identifiers are necessary for any transaction such as a bonded sale: the ID number for the person, and the *erf* number for the property (Figure 8.1).

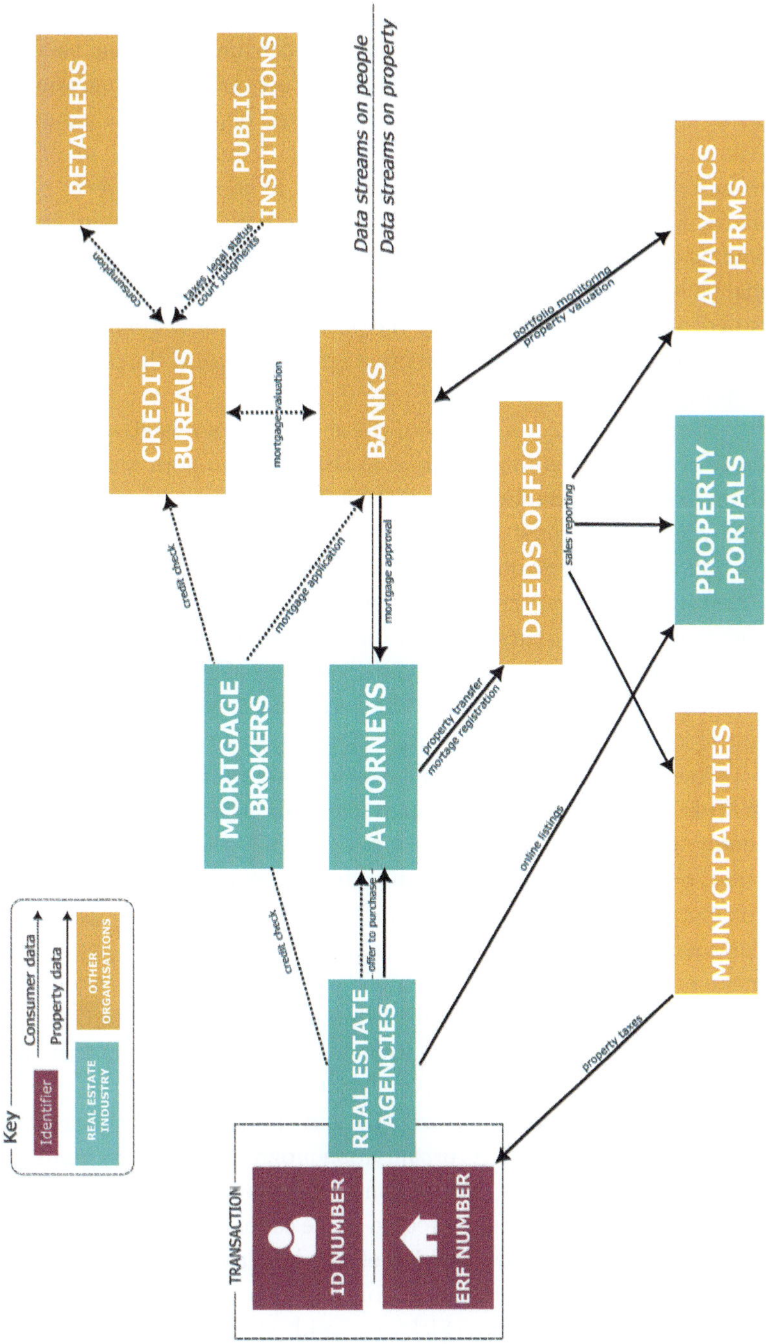

Figure 8.1 The housing market as a flow of data. Source: created by the author.

These identifiers structure informational flows, allowing stakeholders to circulate real-estate and consumer data at various steps in the transaction process: developers use *erf* numbers on printed estate layouts or online sales maps, increasingly popular in the industry; real-estate agents fill in these identifiers on templates such as offers to purchase; mortgage brokers feed them to their software to send mortgage applications to banks; banks use both identifiers to make lending decisions. Attorneys include these identifiers on the documents manually deposited at the Deeds Office, where a new title deed will be issued, marking the finalisation of the sale. A public institution, the Deeds Office manages the national property registry, including the registration of mortgages.

Unpacking the genealogy of these identifiers reveals the *longue durée* of the informational systems in South Africa. Both are rooted in the real-estate techniques of settler colonialism (Rogers 2016), where land was made through maps, cadastral data, deeds and land records (Sud 2020). In the South African 'racial regime of property' (Bhandar 2018), the classification of people into racial categories and bestowed with unequal rights is inseparable from the construction of land markets. The joint categorisation of people and land by property laws enforced the colonial racial order, promoting White economic hegemony through the restriction or denial of property rights for indigenous populations.

Let us examine the origin of the *erf* number: *erf* signifies 'plot of land' in Afrikaans – a clear legacy of Dutch colonisation. From the seventeenth century onwards, the use of cadastral maps and title deeds placed land into the hands of White settlers, allowing the emergence of the landed colonial elite (Dooling 2005). Land information systems were gradually harmonised during the colonial conquest, systematising dispossession in the process: from 1813, no transactions could be authorised without a prior inventory produced by the White-controlled Surveyor General. After the 1913 Native Land Act left only 7 per cent of the land to native populations, the 1927 Land Survey Act standardised the record system across the Union of South Africa, with a unique identifier attributed to each *erf* (Migozzi 2024a). As urbanisation progressed, more *erf* numbers got punched, shaping the informational landscape of housing markets.

ID numbers are similarly rooted in the country's colonial system of population classification, which largely predates apartheid (Christopher 2002): the use of racial categories for segregation purposes dates back to the nineteenth century (Bickford-Smith et al. 1998). Authorities conducted censuses that sorted South Africans into hierarchical, changing and contradictory racial categories, whose only consistency was to erect whiteness at the top of the racialised social order. These techniques of classification paved the way for the apartheid state's mania for measurement (Posel 2000) and love of computers. The Population Registration Act of 1950 required every person to register and obtain an identity card. Deployed from 1955 and distributed to 6.5 million individuals by 1970, ID cards constituted both a 'totem' and an 'instrument' (Breckenridge 2014, 228) of apartheid – with one digit indicating the race of the card holder. To develop centralised databases of population registries, the state contracted with IBM and UK-based ICL.

The importing of computers and the storage of millions of paper-based fingerprints fuelled a managerial, racist ideology of perfect social control and complete segregation. Computers symbolised a modernist technopolitics of identity (Edwards and Hecht 2010): the Minister of Information declared in 1976 that computers made it 'possible for comprehensive data on each individual to be made available for a great many purposes' (quoted in Edwards and Hecht 2010, 627). A 1982 report entitled *Automating Apartheid* condemned the 'data systems [that] make up apartheid's automatic memory bank' (America Friends Service Committee 1982, 14). While incomplete, cumbersome and dysfunctional, these informational infrastructures survived the transition to democracy, and enabled the pioneering rise of the biometric state in South Africa. A 'smart ID programme' was rolled out, issuing new ID numbers with the purpose of not just identifying welfare recipients, but also enforcing surveillance and control (Breckenridge 2014).

The combination of *erf* and ID numbers shaped the production of urban hierarchies and spatial fragmentation. The Group Areas Act of 1951 divided cities into exclusive racial areas (Western 1981): all property transfers were governed by the ideology of racial homogeneity between the racial category of the buyer and the racial category of the property. This state-enforced, race-based fragmentation of the market resulted in extremely unequal market outcomes in terms of residential location, property rights and housing tenure, generating long-lasting inequalities of housing, wealth and life opportunities. Denied property rights in urban areas, Black South Africans were segregated into 'remote locations at the periphery of the city' (Maharaj 2020, 43), placed as tenants of the state housing stock in townships, or left with no option but to reside in informal housing. If democracy erased this racist control of the market, the infrastructural skeleton and identifiers of the apartheid era (Migozzi 2022) underpin contemporary informational flows through the circulation of real-estate and consumer data.

Data streams: assetising traces of debt and deeds

Going back to the visual model (Figure 8.1), I now examine the data streams linked to these two identifiers that condition a bonded sale. Real-estate agents, mortgage brokers and banks source large volumes of data, extracted from the credit reporting system via ID numbers, or resulting from the computerisation of title deeds. Follow the data reveals how the explosion of consumer credit and the rise of real-estate platforms precipitated the unprecedented datafication of people and properties (Migozzi 2024a), allowing credit bureaus and analytics firms to turn data into an economic asset.

When a person is interested in a property, the real-estate agent first needs to make sure the property can be legally sold to the aspiring buyer. This forms part of the 'Know Your Client' requirements under the Financial Intelligence Centre Act, which systematised queries sent to the databases of the Home Affairs National Identification System. Using *erf* and ID numbers, agents can verify the identity of customers and screen them against fraud databases, or check that the seller is also the rightful owner of the property. But identification is not the information that matters the most to them: in a landscape of structural indebtedness and low rates of mortgage approval, agents are mostly interested in accessing information on the customer's financial situation to assess the likelihood of their obtaining a mortgage. To that end, they perform, as early as possible, a 'credit check': before investing time and resource in preparing mortgage applications, real-estate agents, after obtaining consent from the buyer, use ID numbers to source automated credit reports and scores.

The flow of consumer data that produce these reports is vast and always-on, and constitutes an economic asset for credit bureaus. As required by the National Credit Act of 2005, each credit application, each opening of a credit account, each late, missed or completed payment leaves a digital trace circulated through the Data Transmission Hub, which is co-owned by the South African Credit & Risk Reporting Association and the Credit Bureau Association. This reporting system connects a vast amount of data sites, mostly places of consumption (retail stores, supermarkets, car dealers, telecom providers etc.) and places of financial services (credit providers of all sorts, from retails banks to micro-finance companies).

Consumer data are ultimately harvested, and, crucially, enclosed and assetised by major credit bureaus such as TransUnion and Experian. When entering these firms, data are cleaned, enriched, stored and analysed through proprietary algorithms. Bureaus then secure regular and ample income streams by selling scoring solutions or batches of raw data to their clients and business partners. The volume and depth of data they hold is truly enormous – and value is in the volume. In 2021, the National Credit Regulator counted 26.42 million active credit consumers owning more than 85 million credit accounts – twice the size of South Africa's adult population. This information dragnet (Fourcade and Healy 2017) is continuously fed: the third quarter of 2021 counted more than 9.94 million credit applications, while banks and financial institutions generated 271 million inquiries (National Credit Regulator 2021). Through ID numbers, credit bureaus also source biographical and financial data from other institutions: legal information from the court system, property and mortgage data from third-party brokers, and tax data from the South African Revenue Service. The number of people falling under the bureaus' data gaze is expanding through daily consumption: in 2021, 750,000 consumers opened their first credit product in 2021, most of them for a clothing account or a personal loan (TransUnion 2023, 14).

The other stream linked to a bonded sale circulates property data (Figure 8.1). Banks are not only interested in screening the mortgage applicant: the location and value of the property, along with the bank's existing exposure to the neighbourhood, are also critical information that guides lending decisions. Going beyond the scope of their own mortgage databases, banks source geolocated and longitudinal data on past sales and bond registrations, which are registered at the Deeds Office. The circulation of property data from the Deeds Office constitutes another key crossing point for value creation: depending on its directionality, property data either enters a cycle of assetisation, or is integrated into mechanisms of urban governance. In the former case, property data constitute a flourishing business: analytics firms, such as Lightstone, MSCI and Loom, either directly or through data brokers, purchase on a weekly basis the fresh batches of data from the Deeds Office. Data are then cleaned, merged and commercialised to real-estate agencies and financial firms through fee-based platforms. Querying databases with ID numbers or *erf* identifiers, real-estate agents purchase automated reports computed at various scales: the property, the neighbourhood, the suburb or the municipality. The information is rich: aerial imageries, history of transactions (with names and IDs of past and current homeowners), estimated tax rates, size of the plot, surface of the dwelling, number of rooms, distance to various amenities, social composition of the neighbourhood etc. In parallel, public institutions also collect data from the Deeds Office for the purposes of public governance. Municipalities such as the City of Cape Town update their property records in order to calculate property taxes. On a national level, the statistical government office sources data purchased by the South African Reserve Bank from a third-party broker to calculate the Residential Property Price Index, used to inform monetary and macro-economic policies, such as the determination of interest rates.

These two data streams, and the economy they sustain, highlight how the housing market now operates as a stack (Bratton 2016) that performs multiple steps of computation and identification for real-estate actors. Two metrics in particular are required for any bonded transaction: property values and credit scores. These outputs, through which people and houses are converted into digital data, both reflect and re-upload urban inequalities. They exemplify how informational flows, as interpreted by actors, re-enact the differential valuation of human lives and urban contexts.

Data outputs: performing credit scores and property values

As people and properties are converted into digital data, they simultaneously become the subject of automated calculation: during the transaction process, the computational power of informational flows turns people into credit scores, and homes into property values. Following data streams and their outputs reveals here the performativity of informational flows: banks and real-estate agents see homeseekers and properties through the numbers and ordinal categories (high risk/low risk; high value/low value) produced by scoring algorithms. As used and interpreted by real-estate agents and banks to make lending decisions, the combination of credit scores and property values results in highly unequal market outcomes that affect access to and the cost of housing, thereby shaping urban inequalities.

Property values are central to the political economy of housing, and under digital capitalism these values are more visible than ever in South Africa: sales trends and prices are displayed on property portals, while property reports are relatively cheap to buy. But this visibility contrasts with the black-boxed, multi-sited, conflictual determination of property values, which reflects unequal relations of power and capital. The calculation of property values indeed mirrors and enforces the racialised social ordering of people and spaces (Zaimi 2020), as exemplified by the undervaluation of Black neighbourhoods in the USA (Taylor 2019) or by the redlining of South African townships (Kotze and Van Huyssteen 1991).

The calculation of property values profoundly frames any transaction process: property values differentiate buyers in terms of affordability and locational choices, determine profit margins in the industry, and influence municipal budgets. All stakeholders are involved in their determination: developers negotiate selling prices with banks prior to construction and sale in order to secure the participation of mortgages lenders. Banks keep a close eye on property values, contracting with analytics firms to evaluate their mortgage portfolio – a necessary task to meet legal requirements and inform future commercial targets. This evaluation blends in longitudinal and geolocated data: risk-averse banks tend to limit their portfolio exposure in low-income neighbourhoods and townships where repossession is historically difficult (Oldfield and Stokke 2006). Finally, property taxes, which account for 25 per cent of the municipal budget in Cape Town, are determined by property values. The Valuation Department calculates 'market-based' values for each of the 885,018 properties located in the metro area, a process automated through a Computer-Assisted Mass Appraisal system adopted in the late 1990s. Municipal valuations are subject to frequent contestation from dissatisfied homeowners seeking a tax rebate. At every release of the valuation roll, real-estate agents strategically offer services of 'free valuation', contrasting the municipality's 'inaccurate' method with their own expertise.

The other data streams yield credit scores, which are used on a daily basis by real-estate agents, mortgage brokers and banks to process mortgage applications. The consultation of credit scores is another key crossing point where informational flows generate stratifying effects. In South Africa, a stable job and a high income – criteria that only 20 per cent of the population meet – are not enough to get a mortgage: homebuyers must also evidence sufficient credit scores and appropriate financial metrics to mortgage lenders. All South African banks use credit scoring to screen applications and price their mortgage offers, feeding consumer data into their proprietary scorecards (Migozzi 2020). A negative digital trace, such as a missed payment or a court judgment, automatically lowers the score: low-scoring applicants are then flagged as high risk. Credit scoring, therefore, produces unequal market outcomes. A low score is likely to result in outright rejection or higher interest rates, effectively prohibiting homeownership altogether or making it prohibitively expensive. Conversely, a high credit score grants lower rates and more mortgage offers. Most importantly, banks' scorecards and pricing models combine credit scores with loan-to-value, restricting the range of mortgage options available to homebuyers: a low score means that buyers will be asked to increase their deposit, which automatically eliminates a lot of cash-starved, asset-deprived households. The calculation of property values and credit scores therefore discriminates against buyers not just in terms of access to and the cost of housing, but also in terms of the way they are treated during the transaction process. Real-estate agents interpret credit scores on the grounds of moral values, choosing to prioritise or turn away homeseekers. Bad debt, as 'picked up' by agents in credit reports, is associated with unhealthy, immature or irresponsible behaviours.

The hegemonic use of an automated, black-boxed, fee-based technology such as credit scoring meets all legal requirements in South Africa, notably the interdiction of racial discrimination. But if credit scoring is presumably colour-blind, debt is 'startlingly racialised' (James 2018, 818): credit scores – and the data streams that produce them – bear the watermark of previous racial regimes of property and mass indebtedness. Bad debt is prevalent in South Africa: 38 per cent of consumers have impaired records (National Credit Regulator 2022), and 67 per cent of over-indebted consumers are Black African (National Credit Regulator 2017). With stagnating salaries, inflation, mass unemployment (28 per cent) and high poverty rates (48 per cent), the use of credit scores effectively maintains racialised lending patterns, enacting data-driven, acute social sorting on proprietary grounds: scoring algorithms are protected by industrial secrecy. This data output might convey an impression of rationality through the use of statistical modelling (Fourcade and Healy 2024), but it leaves, problematically, unquestioned inaccurate data entries, misattributions of information and calculation errors, and these are common across credit reporting systems (Friedline et al. 2023). As seen by algorithms and their standardised outputs, urban marginality translates into low credit scores, placing the low-ranked population into the algorithmic periphery.

Incalculable, invisible and disqualified: the algorithmic periphery

Under the market's algorithmic gaze, people living at the margin of the digitised economy, working in informal jobs or facing precarious financial situations, are ranked at the bottom of data-driven categories, cast in the algorithmic periphery as invisible, incalculable or disqualified. Similarly, properties located in low-income neighbourhoods such as townships or informal settlements, or associated with other modes of land ownership foreign to the Western definition of property rights, fall into informational gaps and territorial stigma. In that regard, the platformisation of housing markets, through voluminous and granular data streams, uploads 'twentieth century real estate ideologies into twenty-first century information technologies' (Fields and Rogers 2021).

In South Africa, informational flows cover about 6.8 million residential properties, with an estimated total value of 6.7 trillion rand in 2023 (Lightstone 2023). But as wide as this data stream can get, it remains by design remarkably incomplete, failing to take into account the reality of housing situations. Informal settlements, which house 20 per cent of households across the main metropolitan areas (South African Cities Network 2022), remain excluded from the market's digital view: this means that one urban dweller out of five lives in an 'incalculable' house, invisibilised by market reports and data analytics, hidden by property portals, relocated in the informational periphery. The 437 informal settlements and their six hundred thousand Capetonians (Cinnamon and Noth 2023) are therefore treated as digital outcasts. In parallel, digital enrolment and visibility are not synonymous with equal algorithmic treatment: if properties located in Black and Coloured townships are now fully integrated into the market's stack, in stark contrast with the apartheid era, properties in these neighbourhoods remain under-valued compared to formerly White-only areas (Migozzi 2020). The industrial practices, ideologies, norms and datasets that preside over the calculation of property values therefore reproduce the differential valuation of human life enforced by settler colonialism (Dorries et al. 2022).

In contrast, data streams linked to credit scoring bridge the gap between formal and informal housing markets, to encompass virtually the entire population. Whether or not people are aware of credit scoring or willing to be measured, credit scores can be computed and low scores attributed, even in the absence of data: low scores are a necessary output of the statistical modelling. Yet low scores are not only an output of having too much debt, missing payments or being blacklisted: people situated outside the data landscape of credit bureaus are also very likely to be assigned a low score. These people are, in the industrial jargon, 'invisible' customers. Unbanked individuals who pay in cash and avoid registered credit providers are cast into an informational gap hollowed out by the absence of digital history, or by a lower engagement with digitised public institutions. Yet from a statistical viewpoint, data paucity does not stop the calculative possibility: scoring algorithms simply attribute a negative weight to the absence of data, hence associating poor visibility with higher financial risk – and penalising homeseekers for being seen as such. For that reason, agents and mortgage brokers encourage people to 'build up' their credit scores by opening a credit card and using credit parsimoniously, paying back in time. Expanded by new credit users, the data landscape is further extended by credit bureaus and data brokers that seek to harvest alternative data and develop new scoring methods. TransUnion recently launched a new product aptly entitled CreditVision Link, which pretends to 'see millions of previously "credit invisible" and "thin file" South Africans' (TransUnion 2021). Like Experian, the bureau partnered with Chenosis, a 'cross-industry API marketplace', to leverage telecommunications records with the aim of scoring the 'unscoreable' individual (Ranjan 2023).

People with low scores are grouped into two overlapping yet different categories: customers with 'bad debt'; and 'thin-file' or 'invisible' customers. Under the market's algorithmic gaze, being categorised as 'bad debt' or 'invisible' by informational flows leads to 'disqualification' from the transaction process. Depending on their credit scores and income, as well as their personal capacity to negotiate (for instance through private banking), homebuyers are either 'qualified' – that is, rendered eligible for mortgages – or 'disqualified' – categorised as uncreditworthy and rejected by mortgage lenders. As I observed while shadowing agents during fieldwork, banks, which have closed many local branches and reduced in-person interactions, communicate their decisions via email. A minimalist PDF letter states that the application is 'declined on score', 'does not meet the minimum requirements of our credit scorecard', is 'unsuccessful' because of the 'current credit profile', or is declined in view of a 'mitigated risk area'. No further explanation is usually provided. An output of automated calculation, the disqualification of homebuyers interrupts the transaction process. This has wider repercussions in terms of organisational practices: in the context of large-scale indebtedness, the real-estate industry (especially mortgage brokers) encourages homebuyers to obtain 'pre-qualification' certificates to present to the real-estate agent or a developer during initial interactions. This procedure can be completed online with ID numbers. While such a document is worthless from a bank's perspective, the procedure has gained ground in recent years, with the effect of further expanding the information dragnet. When submitting their personal and financial details to obtain a certificate of pre-qualification, aspiring buyers leave other digital traces and enrol themselves in the broker's databases.

To be 'qualified' or 'pre-qualified' under digital ID capitalism (Hicks 2020), homebuyers in need of housing finance are required to be identified as active consumers and to display appropriate, standardised metrics of creditworthiness. Who gets to regulate this performative technology? Few people are even aware of their credit scores in South Africa, despite marketing campaigns to 'be smart about your credit' (Bendall 2023) and recent industrial attempts to penetrate subjectivities by gamifying scores and making them more familiar through smartphone apps. For instance, through a partnership with Experian, Standard Bank's app offers an add-on that provides monthly updates and tips on improving scores. Operating from behind computational walls, away from the buyer's eyes, concentrating data and power, credit bureaus and mortgage lenders score at a distance, exerting their classificatory power under poor public oversight. Neighbourhoods and populations subject to territorial stigmatisation and economic marginality have never been better seen than through an avalanche of data – but informational flows, in terms of metrics and social performativity, recast them in the algorithmic periphery, while turning data into an economic asset.

Widening the scope beyond logistics and exclusion

Seen from the digitised housing sphere, the making of informational peripheries needs to be captured from a double logic of *integration* (incorporating people and places into networked databases) and *ordering* (computing and interpreting unequal categories to read and act upon the social world). In that regard, the method of 'follow the data' reveals the inherent incompleteness and fragmented nature of data, and the performative function of identification systems. Yet the process of social ordering by identification, and the unequal networked geographies it sustains and creates, are not sufficiently acknowledged as a productive force in the current conceptualisation of informational flows and peripheries. One way to solve this tension would be to increase the empirical focus beyond logistics, and interpret the 'infrastructural and technological redlining' provoked by informational flows (Datta 2024) not in terms of 'bypassing' and 'exclusion', but rather as a process of unequal inclusion.

If logistics is key to the unequal incorporation of territories into global production networks, the new regime of governance that defines informational peripheries extends beyond the realm of logistics, as evidenced by the global spread of credit scoring and identification systems within and beyond emerging economies (Curran and Smart 2021; Hicks 2020; Opalo 2022; Wong 2003). Informational peripheries surge through the mediums of warehouses and labour markets, and also through the ideology of identification and measurement – most of the time promoted by the alliance between the digitised state and IT companies, under the banner of promoting financial inclusion and fighting fraud and terrorism. E-commerce and social scoring are of course deeply related: as labour, trade, financial services and consumption become digitally mediated, more digital traces are generated, advancing the enrolment of people and territories into data regimes. Informational flows enact two logics: ordering by differentiation, given the uneven meshes of the information dragnet and the ordinal categories they produce; and integration by identification, given the organisational imperative to increase the volume and granularity of data.

These logics, weaving 'social measurement, differentiation, and hierarchy' (Fourcade 2021) through data-driven identification systems, therefore question what is currently designated as 'technological and infrastructural redlining' in the current concept of the informational periphery. At present, redlining designates how informational flows generate processes of bypassing and exclusion as productive mechanisms of the informational periphery (Datta 2024). Yet redlining is not only a practice of place-based exclusion, but also a form of discriminatory inclusion – seeking profit by charging people in poor areas higher rates for loans, mortgages or insurance (Aalbers, 2005). As an increasingly 'data-driven' practice (Koopman 2022), redlining should, then, rather be understood as unequal inclusion, computing preferential treatment or predatory lending practices (Taylor 2019; Wyly et al. 2009). Informational flows, especially at the scale of individuals, enact a shift towards individual-based treatment, charting unequal paths to financial services or housing. Current patterns of technological and infrastructural plumbing, as driven by state authorities, mean that more and more people are actually included in, rather than excluded from, informational flows – even if such inclusion takes place on asymmetrical, algorithmically obscure grounds (Burrell and Fourcade 2021). People might be 'fungible' for logistics labour (Datta 2024) in the informational periphery, but also, crucially, they are fungible for the purposes of market making and political governance, identifiable and calculable in a myriad of ways for political and commercial goals. It seems necessary to investigate the full spectrum of inclusion under technological redlining (Noble 2018) to fully grasp the making of informational peripheries.

Follow the data, find the digital outcasts

In this chapter I adopted 'follow the data' as a methodology, emphasising the need to define a sectoral scope, a starting point and a directionality to investigate data flows across infrastructures and organisations, to detect key crossing points of enclosure and assetisation, and to reveal when algorithms and data outputs become performative over urban lives and spaces. Informed by in-depth fieldwork within the real-estate industry in South Africa, I applied this method to map and model data flows across the housing market, choosing the generic scenario of a bonded transaction, and following data upstream and downstream from this event.

Harvesting traces of debt and streams of deeds, the real-estate industry feeds and plugs into always-on informational flows, through which people and properties are enrolled into data regimes and converted into digital data. The credit reporting system and the digitisation of title deeds feed two streams of data, both subject to enclosure and assetisation: as consumer and property data enter the proprietary grounds of credit bureaus and analytics firms, they become an economic asset – a resource and a property that generates revenue streams (Birch and Ward 2024). The calculative output of these streams – property values and credit scores – is commercialised to real-estate professionals, who interpret them to sort customers and allow or deny access to housing finance.

This chapter underlined how informational flows introduce a new regime of algorithmic visibility, in which people and properties 'must be visible, and available for measurement' (Fourcade and Healy 2017, 19). People and houses falling into informational gaps, failing to meet appropriate thresholds due to the structure of the economy, are ranked in the algorithmic periphery – categorised as low scoring, invisible or incalculable. Houses located in stigmatised neighbourhoods are subjects to under-valuation, while precarious and indebted homebuyers, wherever they live, can be disqualified from the transaction process. In that regard, the algorithmic periphery of post-apartheid South Africa attests that digital technologies have indeed broken down the 'conventional conflation of urban edge with the urban periphery' (Datta 2024). Yet it also highlights how social ordering in the digital world remains deeply connected to effects in the material world, as racialised inequalities of debt and income are uploaded into informational flows and affect their output.

Following the data, this chapter underlined the *longue durée* of infrastructures and identifiers, such as ID and *erf* numbers, through which 'human identities' and properties are converted into digital data

(Masiero 2023). Created to exclude, townships were not thrown into informational oblivion, but placed under the authoritarian gaze of the central state, municipal authorities and police forces. A dominated periphery, townships were imperfectly surveyed but violently surveilled. Property regimes obeyed a for-profit, racialised politics of identity, with identification systems paving the way for the datafication of people and property in the digital era. Pushing towards on-demand, granular information, real-estate stakeholders stack up financial and biographical data on properties and people: this networking of the digital and material worlds perpetuates the differential valuation of urban lives, casting the precarious and the informal into the informational periphery. These patterns of digitisation threaten to produce a poisonous equation between the algorithmically unseen, the economically informal and the politically illegitimate.

References

Aalbers, M. B. 2005. Place-based social exclusion: Redlining in the Netherlands. *Area* 37(1): 100–9. https://doi.org/10.1111/j.1475-4762.2005.00609.x.

Akbari, A. 2020. Follow the thing: Data. *Antipode* 52(2): 408–29. https://doi.org/10.1111/anti.12596.

America Friends Service Committee. 1982. *Automating Apartheid: U.S. computer exports to South Africa and the arms embargo*. NARMIC/America Friends Service Committee.

Ash, J., Kitchin, R. and Leszczynski, A. 2024. *Researching Digital Life: Orientations, methods and practice*. Thousand Oaks, CA: Sage Publications.

Bates, J., Lin, Y-W. and Goodale, P. 2016. Data journeys: Capturing the socio-material constitution of data objects and flows. *Big Data & Society* 3(2). https://doi.org/10.1177/2053951716654502.

Bendall, B. 2023. Be smart about credit. Bizcommunity, 24 October. Accessed February 2025. https://www.bizcommunity.com/Article/196/513/243047.html.

Bhandar, B. 2018. *Colonial Lives of Property: Law, land, and racial regimes of ownership*. Durham, NC: Duke University Press.

Bickford-Smith, V., Van Heyningen, E. and Worden, N. 1998. *Cape Town: The making of a city*. Cape Town: David Phillip Publishers.

Birch, K. 2023. *Data Enclaves*. London: Palgrave Macmillan.

Birch, K., Chiappetta, M. and Artyushina, A. 2020. The problem of innovation in technoscientific capitalism: Data rentiership and the policy implications of turning personal digital data into a private asset. *Policy Studies* 41(5): 468–87. https://doi.org/10.1080/01442872.2020.1748264.

Birch, K. and Ward, C. 2024. Assetization and the 'new asset geographies'. *Dialogues in Human Geography* 14(1): 9–29. https://doi.org/10.1177/20438206221130807.

Bratton, B. H. 2016. *The Stack: On software and sovereignty*. Cambridge, MA: MIT Press.

Breckenridge, K. 2014. *Biometric State: The global politics of identification and surveillance in South Africa, 1850 to the present*. Cambridge: Cambridge University Press.

Burrell, J. and Fourcade, M. 2021. The society of algorithms. *Annual Review of Sociology* 47(1): 213–37. https://doi.org/10.1146/annurev-soc-090820-020800.

Chorley, R. J. and Haggett, P. (eds.). 1967. *Models in Geography*. London: Methuen.

Christopher, A. J. 2002. 'To define the indefinable': Population classification and the census in South Africa. *Area* 34(4): 401–8. https://doi.org/10.1111/1475-4762.00097.

Christophers, B. 2011. Follow the thing: Money. *Environment and Planning D: Society and space* 29(6): 1,068–84. https://doi.org/10.1068/d8410.

Cinnamon, J. and Noth, T. 2023. Spatiotemporal development of informal settlements in Cape Town, 2000 to 2020: An open data approach. *Habitat International* 133. https://doi.org/10.1016/j.habitatint.2023.102753.

Cook, I. 2004. Follow the thing: Papaya. *Antipode* 36(4): 642–64. https://doi.org/10.1111/j.1467-8330.2004.00441.x.

Cowan, T. 2021. Uncertain grounds: Cartographic negotiation and digitized property on the urban frontier. *International Journal of Urban and Regional Research* 45(3): 442–57. https://doi.org/10.1111/1468-2427.13016.

Curran, D. and Smart, A. 2021. Data-driven governance, smart urbanism and risk-class inequalities: Security and social credit in China. *Urban Studies* 58(3): 487–506. https://doi.org/10.1177/0042098020927855.

Dalton, C. M., Taylor, L. and Thatcher, J. 2016. Critical data studies: A dialog on data and space. *Big Data & Society* 3(1). https://doi.org/10.1177/2053951716648346.

Datta, A. 2024. The informational periphery: Territory, logistics and people in the margins of a digital age. *Asian Geographer* 41(2): 125–42. https://doi.org/10.1080/10225706.2023.2253233.

Dooling, W. 2005. The making of a colonial elite: Property, family and landed stability in the Cape Colony, c.1750–1834. *Journal of Southern African Studies* 31(1): 147–62. https://doi.org/10.1080/03057070500035802.

Dorries, H., Hugill, D. and Tomiak, J. 2022. Racial capitalism and the production of settler colonial cities. *Geoforum* 132: 263–70. https://doi.org/10.1016/j.geoforum.2019.07.016.

Edwards, P. N. and Hecht, G. 2010. History and the technopolitics of identity: The case of Apartheid South Africa. *Journal of Southern African Studies* 36(3): 619–39. https://doi.org/10.1080/03057070.2010.507568.

Fields, D. 2024. Digital experiments with landed property: Robots, race, and rent. *Tijdschrift voor economische en sociale geografie* 115(3): 329–45. https://doi.org/10.1111/tesg.12630.

Fields, D. and Rogers, D. 2021. Towards a critical housing studies research agenda on platform real estate. *Housing, Theory and Society* 38(1): 72–94. https://doi.org/10.1080/14036096.2019.1670724.

Flensburg, S. and Lai, S. S. 2023. Follow the data! A strategy for tracing infrastructural power. *Media and Communication* 11(2): 319–29. https://doi.org/10.17645/mac.v11i2.6464.

Fourcade, M. 2021. Ordinal citizenship. *British Journal of Sociology* 72(2): 154–73. https://doi.org/10.1111/1468-4446.12839.

Fourcade, M. and Healy, K. 2017. Seeing like a market. *Socio-Economic Review* 15(1): 9–29. https://doi.org/10.1093/ser/mww033.

Fourcade, M. and Healy, K. 2024. *The Ordinal Society*. Cambridge, MA: Harvard University Press.

Friedline, T., Stewart, K., Bolinger, C. and Wood, A. K. 2023. Credit scoring as a carceral practice: An abolitionist framework. *Race and Social Problems* 16: 230–48. https://doi.org/10.1007/s12552-023-09406-6.

Galster, G. C. 2023. How digitalisation influences neighbourhood change. *Urban Studies* 61(16): 3028–49. https://doi.org/10.1177/00420980231198197.

Graham, S. and Marvin, S. 2001. *Splintering Urbanism: Networked infrastructures, technological mobilities and the urban condition*. Abingdon: Routledge.

Hicks, J. 2020. Digital ID capitalism: How emerging economies are re-inventing digital capitalism. *Contemporary Politics* 26(3): 330–50. https://doi.org/10.1080/13569775.2020.1751377.

Hughes-McLure, S. 2022. Follow the money. *Environment and Planning A: Economy and space* 54(7): 1299–322. https://doi.org/10.1177/0308518X221103267.

James, D. 2014. *Money from Nothing: Indebtedness and aspiration in South Africa*. Redwood City, CA: Stanford University Press.

James, D. 2018. Mediating indebtedness in South Africa. *Ethnos* 83(5): 814–31. https://doi.org/10.1080/00141844.2017.1362450.

Kass, A. 2020. Working with financial data as a critical geographer. *Geographical Review* 110(1–2): 104–16. https://doi.org/10.1080/00167428.2019.1684193.

Kitchin, R. 2013. Big data and human geography: Opportunities, challenges and risks. *Dialogues in Human Geography* 3(3): 262–7. https://doi.org/10.1177/2043820613513388.

Kitchin, R. 2014. *The Data Revolution: Big data, open data, data infrastructures and their consequences*. Thousand Oaks, CA: Sage Publications.

Koopman, C. 2022. The political theory of data: Institutions, algorithms and formats in racial redlining. *Political Theory* 50(2): 337–61. https://doi.org/10.1177/00905917211027835.

Kotze, N. J. and Van Huyssteen, M. K. R. 1991. Rooi-omlyning in die behuisingsmark van Kaapstad [Redlining in the housing market of Cape Town]. *South African Geographer/Suid-Afrikaanse Geograaf* 18(1–2): 97–122.

Lightstone. 2023. *Property Newsletter*, September 2023. Accessed February 2025. https://www.lightstone.co.za/newsletters/property/2023/Sept/A1Sep23.

Maharaj, B. 2020. The apartheid city. In *Urban Geography in South Africa: Perspectives and theory*, edited by R. Massey and A. Gunter, 39–54. Dordrecht: Springer.

Masiero, S. 2023. Digital identity as platform-mediated surveillance. *Big Data & Society* 10(1). https://doi.org/10.1177/20539517221135176.

Migozzi, J. 2020. Selecting spaces, classifying people: The financialization of housing in the South African city. *Housing Policy Debate* 30(4): 640–60. https://doi.org/10.1080/10511482.2019.1684335.

Migozzi, J. 2022. Apartheid by algorithm. *Logic(s)* 17. Accessed February 2025. https://logicmag.io/home/apartheid-by-algorithm.

Migozzi, J. 2024a. The good, the bad and the tenant: Rental platforms renewing racial capitalism in the post-apartheid housing market. *Environment and Planning D: Society and space* 42(4): 534–58. https://doi.org/10.1177/02637758231195962.

Migozzi, J. 2024b. Maps, apps and race: The market as a theoretical machine. *Tijdschrift voor economische en sociale geografie* 115(3): 346–52. https://doi.org/10.1111/tesg.12631.

National Credit Regulator. 2017. Over-indebtedness: National Treasury & National Credit Regulator. Parliamentary Monitoring Group, 14 June. Accessed February 2025. https://pmg.org.za/committee-meeting/24602.

National Credit Regulator. 2021. Credit Bureau Monitor, Third Quarter – September 2021. Accessed February 2025. https://www.ncr.org.za/documents/CBM/CBM%20Q3%202021.pdf.

National Credit Regulator. 2022. Credit Bureau Monitor, Fourth Quarter – December 2022. Accessed February 2025. https://www.ncr.org.za/documents/CBM/CBM%20Q4%202022.pdf.

Nethercote, M. 2023. Platform landlords: Renters, personal data and new digital footholds of urban control. *Digital Geography and Society* 5. https://doi.org/10.1016/j.diggeo.2023.100060.

Noble, S. U. 2018. *Algorithms of Oppression: How search engines reinforce racism*. New York: New York University Press.

Oldfield, S. and Stokke, K. 2006. Building unity in diversity: Social movement activism in the Western Cape anti-eviction campaign. In *Voices of Protest: Social movements in post-apartheid South Africa*, edited by R. Ballard, A. Habib and I. Valodia, 111–32. Natal: University of KwaZulu/Natal Press.

Opalo, V. W. 2022. Credible risk: Private credit bureaus and the work of loan officers in West Africa. *Africa* 92(4): 625–43. https://doi.org/10.1017/S0001972022000353.

Open Data Institute. 2020. What is an identifier? Accessed February 2025. https://theodi.org/insights/explainers/explainer-what-is-an-identifier.

Pollio, A. and Cirolia, L. R. 2022. Fintech urbanism in the startup capital of Africa. *Journal of Cultural Economy* 15(4): 508–23. https://doi.org/10.1080/17530350.2022.2058058.

Posel, D. 2000. A mania for measurement: Statistics and statecraft in the transition to apartheid. In *Science and Society in Southern Africa*, edited by S. Dubow, 116–42. Manchester: Manchester University Press.

Ranjan, A. 2023. Experian and Chenosis partner to drive financial inclusion through data in South Africa. TechAfrica News, 19 October. Accessed February 2025. https://www.techafricanews.com/2023/10/19/experian-and-chenosis-partner-to-drive-financial-inclusion-through-data-in-south-africa.

Rob, K. 2021. *Data Lives: How data are made and shape our world*. Bristol: Policy Press.

Rogers, D. 2016. *The Geopolitics of Real Estate: Reconfiguring property, capital, and rights*. Lanham, MD: Rowman & Littlefield.

Sadowski, J. 2020. The internet of landlords: Digital platforms and new mechanisms of rentier capitalism. *Antipode* 52(2): 562–80. https://doi.org/10.1111/anti.12595.

Salganik, M. J. 2017. *Bit by Bit: Social research in the digital age*. Princeton, NJ: Princeton University Press.

Sawyer, S., Crowston, K. and Wigand, R. T. 2014. Digital assemblages: Evidence and theorising from the computerisation of the US residential real estate industry. *New Technology, Work and Employment* 29(1): 40–56. https://doi.org/10.1111/ntwe.12020.

South African Cities Network. 2022. *State of South African Cities: Report*. Accessed February 2025. https://www.sacities.net/wp-content/uploads/2023/03/SoCR-V-2021-03.20_WEB.pdf.

Sud, N. 2020. *The Making of Land and the Making of India*. Oxford: Oxford University Press.
Taffel, S. 2023. AirPods and the earth: Digital technologies, planned obsolescence and the Capitalocene. *Environment and Planning E: Nature and space* 6(1): 433–54. https://doi.org/10.1177/25148486221076136.
Taylor, K. Y. 2019. *Race for Profit: How banks and the real estate industry undermined black homeownership*. Durham, NC: University of North Carolina Press.
TransUnion. 2021. New scoring solution gives lenders insights to help 'credit invisible' South Africans. 17 January. Accessed February 2025. https://newsroom.transunion.co.za/new-scoring-solution-gives-lenders-insights-to-help-credit-invisible-south-africans.
TransUnion. 2023. *Empowering Credit Inclusion: A deeper perspective on new-to-credit consumers*. Accessed February 2025. https://www.transunion.com/content/dam/transunion/global/consumer/documents/empowering-credit-inclusion-a-deeper-perspective-on-new-to-credit-consumers.pdf.
Western, J. 1981. *Outcast Cape Town*. 2nd edn. Berkeley, CA: University of California Press.
Wojcik, D., Iliopoulos, P., Ioannou, S., Keenan, L., Migozzi, J., Monteath, T., Pazitka, V., Torrance, M. and Urban, M. 2024. *Atlas of Finance: Mapping the global story of money*. New Haven, CT: Yale University Press.
Wong, D. W. 2003. Spatial decomposition of segregation indices: A framework toward measuring segregation at multiple levels. *Geographical Analysis* 35(3): 179–94. https://doi.org/10.1111/j.1538-4632.2003.tb01109.x.
Wyly, E., Moos, M., Hammel, D. and Kabahizi, E. 2009. Cartographies of race and class: Mapping the class-monopoly rents of American subprime mortgage capital. *International Journal of Urban and Regional Research* 33(2): 332–54. https://doi.org/10.1111/j.1468-2427.2009.00870.x.
Zaimi, R. 2020. Making real estate markets: The co-production of race and property value in early 20th century appraisal science. *Antipode* 52(5): 1539–59. https://doi.org/10.1111/anti.12642.

9

Digitising peripheral property: assetisation, platformisation and disconnection

Thomas Cowan

In late 2020, an online ceremony was held to launch the Svamitva Yojana, a scheme that forms part of the national 'Digital India' campaign to digitally transform government services. Svamitva, which takes its name from the Sanskrit word for 'ownership', aims to map and distribute private ownership titles to over six hundred thousand *abadi* (village residential) territories across the country. This endeavour is supported by the technology, geo-positioning reference stations and mapping platforms that will enable state officials to map private property boundaries on traditional customary lands, distribute and manage private ownership rights, and wire up rural residents and their newly minted assets to digital property platforms (Government of India 2020). The scheme is being overseen by an expert committee that includes the National Informatics Centre, the Survey of India, various state offices, financial institutions and domestic and international tech firms, and provides the base digital infrastructure for a host of property-dependent citizenship and finance schemes. At the event, various government officials, including prime minister Narendra Modi, heralded the remarkable promise of digital public infrastructures and digitised private property titles to unleash the Indian population from the stranglehold of local state and communitarian institutions, and into the warm embrace of digitalised real-estate and finance networks.

Under the scheme, drone mapping provides computer engineers and financial institutions with a direct line of sight into otherwise peripheral and opaque rural territories, enabling more efficient state infrastructural planning, fulsome fiscal recovery and, importantly, the transfer of customary rural shareholdings into digitised private property assets. The scheme principally seeks to recentre peripheral, customary uses and purposes of land into the strict cartographic registers of global property systems. Indeed, the government promises to make historic changes to rural India by allowing residents to utilise their rural landholdings as financial assets, providing access to financial credit that, echoing the much-discredited neoliberal orthodoxies of Hernando de Soto (2000) and C. K. Prahalad (2004), will eradicate poverty and supercharge socio-economic development.

At the opening ceremony, held over video-link during the COVID-19 pandemic, a beneficiary of one of the first digitally integrated property cards issued under the scheme was invited to speak about the impact. The middle-aged man, a mason from Yamunanagar village in the state of Haryana, echoed the narrative of digitalised financial development at the centre of the scheme:

> Before the launch of the scheme, when a person started construction work of his house in the village, a dispute arose among people due to the absence of any kind of ownership records … But this system will put a check on disputes … [now] I have got the right to ownership of my house. A bank has invited me to issue a loan of Rs 3 lakh to me. With this amount of loan, I will start a new work of shuttering plates, which will help me to grow in life. (Kumar Sharma, 2020)

The mason's comments echo those of central government, and its international partners, on the remarkable power of digitally charged land titling schemes like Svamitva not only to radically reconfigure state–village relations – ushering in a revolution of digitally mediated rural development – but also to 'unleash the productive potential of the countryside' (*The Hindu* 2020), opening up rural land to domestic finance and real-estate markets. Indeed, both the scheme and the tech sector that supports it pose the problems of transferring non-privatised, peripheral property into platformised real estate as a purely computational procedure – one in which the complexities of arbitrating competing property claims and enclosing non-privatised property are brushed aside by the cold rationality of technological vision. Here digital technologies – drones, mapping software, geo-positioning systems, artificial intelligence – are being utilised to establish entirely new ways of seeing, perceiving and mobilising land and property relations, in ways that conceal their underlying socio-technical mediation (see Amoore 2020).

And yet, despite euphoric accounts of the promise of digitisation for India's customary property systems, digitally mediated titling programmes have a far more chequered track record in practice. As numerous studies have shown, progress on a range of land and property digitisation schemes across the country that have been ongoing since the early 1990s has been uneven at best, beset with difficulties in rewiring diverse tenurial and record systems into the clean and partible data categories of an integrated digital registry (Benjamin 2005; Benjamin and Raman 2011; Cowan 2021; Richter 2011; Jonnalagadda et al. 2021; Goswami et al. 2017; Hasan 2024). My time spent observing the work of state officials, surveyors and private consultants working on Svamitva and other state-organised property technology (PropTech) schemes has revealed a set of initiatives plagued by speculation and frailty, mired in disputes over jurisdiction and ownership, fraught with challenges over computation, transference and classification, and subject to all manner of localised malfeasance. As a wealth of literature has argued, property relations are both sticky and capacious, imbued with competing claims, meanings and systems of recognition that are not as easily unpicked and converted as is often suggested within policy literature (Sikor and Lund 2009; Hall et al. 2011; Blomley 2005; Verdery 1999).

Some months after the ceremony, I was discussing the progress of the scheme with a senior government official. After some discussion of the various legal and technical obstacles scheme officials had been navigating since our previous meeting, he noted that the mason from the programme's launch had recently called, asking to withdraw himself from the scheme. The official explained:

> You see the mason, from the scheme launch, he was in touch with us only recently and had asked to remove his property from the system, he didn't want to be on the system, offered his card back. It turned out that once his property was registered, he was no longer eligible for various provisions [government welfare programmes] ... this is why what we are doing is so important.

It turns out that the mason's family were one of eight hundred thousand across the state of Haryana made ineligible for state rations following the incorporation of their property data in the state's algorithmic family identification system, Parivar Pehchan Patra (Siwach 2023). For the official this was evidence of the power of the scheme to remedy corruption within existing public welfare systems and fed into a broader narrative that Svamitva data, digitalised property, when imbricated into larger digital public infrastructures can act as the harbinger of more efficient governance. And yet the story also reveals how India's PropTech revolution is not *just* about consolidating a new asset-class in rural property. Rather, by integrating digitalised property data with digital public infrastructure, India's property digitalisation schemes are rewriting the very character of both property and citizenship, producing new forms of propertied citizenship and peripheralisation.

As I will discuss in this chapter, these new propertied peripheries emerge both through infrastructural 'torque' (Bowker and Star 2000) – the inability of diverse ways of holding, using and understanding property to compute with core Svamitva data categories – and, as the mason found, through the ways in which shifting clusters of data are being used to differently sort, link and unlink people with basic rights and public services (see Phan and Wark 2021). Both are affecting new forms of dispossession and abjection determined by computational factors quite distinct from those of the postcolonial moment. In other words, property digitalisation programmes are as much tools of assetisation and platformisation as they are conduits toward a configuration of new, and perhaps rather unpredictable, form of digital citizenship.

Following the theme of this collection, I conceptualise 'peripheral property' not as a fixed spatial category, but rather as those sites excised from or oppositional to the mainstream, computable grammars of urban real estate. These are the property relations that remain pixellated in large-scale property data visualisations and asterisked in databases, beholden to quite other forms of digital representation, and often present in 'low resolution' (Singh and Jackson 2021) within digital public infrastructures. These peripheries are importantly not simply analogue hinterlands that prefigure certain digitalisation, but rather, we find, to borrow from Buzan and Lawson (2013, 9), 'peripheries in the core and cores in the periphery'. The question of peripherality is not so much about geographical location as it is about a subaltern relation to the mainstream grammars of digital infrastructure (Cowan 2022).

In the rest of the chapter, I first situate informational peripheries in the Global South with reference to existing debates on property technologies that are largely taken from examples in the post-industrial urban Global North. The chapter then considers how property technology's empirically distinct form and deployment in the majority world alters our understanding of processes of land and property digitalisation in the present. Subsequently the chapter explores how state-led PropTech schemes are engaged in overcoming core barriers to capture and reconfigure the periphery, producing propertied forms of informational periphery. These barriers are understood in three key refrains that structure the chapter: assetisation, platformisation and peripheralisation. Together, these constitute an informational infrastructure that is producing what the editors of this collection refer to as 'new configurations of peripheralisation' (Datta and Hoefsloot, Chapter 1 this volume), anchored to and yet producing peripheries anew.

This chapter draws on insights developed during ongoing research undertaken in north India from 2023. This research consists of ethnographic participant observation with government officials, drone surveyors and ICT engineers working on a range of the Government of India's PropTech initiatives in the states of Haryana, Himachal Pradesh and Rajasthan. My ability to conduct participant observation research with government officials has been facilitated by my membership of an elite research institution in the Global North, and research affiliation with an elite research institution in India. My position as a foreign, white male researcher overdetermines my access to a largely masculinised field and the kinds of data I have access to. The recuperation of data by elite Northern research scholars and institutions recreates material imbalances in knowledge production, dissemination and access that serve to stricture the epistemological boundaries of scholarly knowledge, and enables elite and post-imperial nations and subjects such as myself to set the terms and reap the benefits of scholarly pursuit in ways that mirror the traffic of colonial knowledge concerning property and governance that is the subject of this study. This study considers this recursive and yet contingent material and epistemological power relation centrally and is broadly concerned with scrutinising the circulation of elite ideology, structures and practices that sustain conventional power in the former metropole and rising autochthonous nationalism and corporate capitalism in the former periphery. This project takes seriously how these circulations are being organised against and disrupted and how *other* systems of land use and human relation might be organised.

Provincialising PropTech

Debates concerning the increasing digitalisation of urban processes have, in recent years, examined the proliferation of property technologies, or PropTech, increasingly deployed by the state, corporate landlords and the real-estate sector to manage real-estate portfolios (Fields 2022), algorithmically manage housing markets (Safransky 2020; Migozzi 2020), and deepen landlords' surveillance and regulation of tenants

(McElroy and Vergerio 2022). The expanding arsenal of technologies increasingly governed through artificial intelligence and hardwired into land and housing markets is, it is argued, having a profound impact upon the received (metropolitan) geographies of capitalist urbanisation – of gentrification, financialisation, urban policing (Jefferson 2018), housing speculation and dispossession (Safranksy 2020). And yet, PropTech research within urban geography remains, with certain exceptions (Faxon and Wittekind 2023; Cowan 2021; Migozzi et al. 2023; Benjamin and Raman 2011), focused largely on contexts in the post-industrial urban North. Not only do these studies map onto existing imbalances in knowledge production in and of the global urban (Bhan 2019), they also, importantly, draw from contexts largely defined by the total hegemony of private property and the political dominance of mainstream financial and banking institutions.

In much of the world, property relations and with them the object of property technologies are quite different. Property in land in the majority world is composed by a mixture of thick private, public, common and customary land tenures; the afterlives of colonial and socialist land struggle and settlement (Verdery 1999; Hsing 2010; Cowan 2022; Ghertner 2014). As scholars of property in the South have argued, subtending these diverse property regimes are a host of state and non-state institutional apparatuses that draw their legitimacy and sovereignty from their capacity to uphold non-normative forms of holding, owning and using property (Sikor and Lund 2009; Harriss-White 2003). These customary institutions uphold non-privatised property systems which Ghertner has argued can 'nurture forms of collective governance resolutely opposed to privatisation' (Ghertner 2015, 554) and provide forms of inhabitation and economic activity that preclude capital's ability to fully subsume land and housing to market forces. They also, we might add, facilitate circuits of credit that distance urban and rural communities from formal property-leveraged financial institutions (Goodfellow 2020).

In this context, the application of property-based theories from the post-industrial North to the majority world must be attentive to three key distinctions present in the majority world. First, the discrete institutional forms and logics through which both property and social life are governed in the majority world. These forms and logics of rule exceed the notional Weberian institutional apparatuses that we come to associate with property governance in the abstract, and are rather attuned to the contingent, makeshift and often communitarian politics of claim making that arise from the overtures of state developmentalism – that which Chatterjee (2004) refers to as 'political society'. For our purposes this ‘political society’ imbibes property relations with what Solomon Benjamin (2005) refers to as a powerfully fluid set of social, legal and institutional relations through which claims and counter-claims to ownership, use and exclusion are established (Hall et al. 2011).

Secondly and relatedly, as Ghertner notes, our understanding of property must be attentive to the explicitly uneven legacies of land enclosure and privatisation – a context defined by global capital’s incomplete ‘counterrevolution’ against the ‘social uses of land’ that dominate both urban and rural land tenure in the majority world (Ghertner 2015, 553). The persistence of non-privatised property and non-market-oriented institutional logics in the sinews of otherwise fully enclosed and assetised territory together act as central barriers to capital’s real subsumption of land and the extension of financial accumulation therein. Property relations in peri-urban north India, where my fieldwork on PropTech has taken place, are highly diverse, composed of a mixture of commons, coparcenary, collective and private rights. Property records and mutations are managed through disaggregated, tabular and non-linear ledgers and synoptic cadastres, and property rights are presumptive, reliant upon the adjudication of claims and counter-claims manifest in deeds, maps and physical possession (Jonnalagadda and Cowan 2024).

It is precisely this context that affords PropTech an altogether different purpose and quality. If in the urban North Atlantic PropTech is being layered onto established ecosystems of private property and is utilised to consolidate the financialisation of housing markets and deepening avenues for rentier accumulation (Fields 2022; Sadowski 2020; McElroy and Vergerio 2022), then in much of the world, the work of PropTech is far more transformational – used not only to more efficiently manage property assets, but also to swipe away diversity and create property assets anew.

Finally, in this context, the state – in partnership with the private sector – takes a more central role in PropTech initiatives in India than discussed elsewhere in the literature. The various PropTech schemes in India that this chapter discusses are organised under public–private management structures, influenced in part by New Public Management approaches, that emphasise infusing public-sector bodies with private-sector expertise, efficiency and cost-recovery (see Hasan 2022). These public–private partnerships explicitly seek to utilise digital infrastructures to batten down customary barriers to market-oriented socio-economic development, enclose and formalise peripheralised property regimes, and use property data to more efficiently link people to both state and financial services (Hasan 2022).

While there certainly are private-sector real-estate and tech actors utilising property technologies within pockets of privatised urban housing markets for purposes similar to those in the North Atlantic (Migozzi et al. 2023), the sheer size and scale of state-led PropTech programmes dwarfs these private counterparts. These include the Digital India Land Records Modernisation Programme (DILRMP) to digitise land records across the country; the aforementioned Svamitva scheme that aims to enclose and platformise non-privatised land; the Unique Land Parcel Identification Number (ULPIN) scheme to establish an '*Aadhaar* for property', assigning a 14-digit alpha-numeric number to each geo-coordinated landholding, imbricated with personal identification data to allow property to circulate like a fungible resource through public and commercial digital infrastructures; and various experiments in the use of blockchain technologies for property digitisation (see Kshetri 2022).

The qualitatively distinct organisational, ideological and technological purpose of PropTech schemes in places like India situates this technology within an entirely distinct and less well understood context, one prefigured by the enduring diversity of property relations, institutions and logics of rule. Indeed, the Indian government's PropTech adventure brings together core policy commitments to reform local property governance, open up property-based blockages to large-scale urban and infrastructural capital investment, and plug rural populations into emerging financial markets and new digitised infrastructures of citizenship. PropTech in India, as such, is principally a technology of the periphery, one that acts as *the* central instrument for the recalibration of core informational grammars through which both markets and governance take place.

Property (un)settlement and digital salvation

Following the pro-market economic reforms introduced under the New Economic Policy in 1991, and with perhaps greater haste following the deregulation of the real-estate and finance sectors from the mid-2000s, a whole host of spatial planning and development technologies – from special economic zones to mega-infrastructural corridors and smart cities – have been deployed to capture global capital investment in extended geographies of urban and infrastructural development (Bathla 2022; Kanai and Schindler 2022). To this, enduring non-privatised property regimes, situated in the rural and peri-urban spaces touted for capital investment, represent a significant barrier.

The current spate of government property digitisation schemes arrives as something of a silver bullet aimed at eradicating these peripheral barriers, refiguring property in land onto data platforms consolidating new digitally mediated forms of propertied citizenship. These form part of a longer history of government land and property digitisation programmes, which since the early 2000s have explicitly targeted the private titling and formalisation of customary lands (Nayak 2013), often under the purview of creating legibility (Benjamin and Raman 2011). The challenge for PropTech engineers, planners and officials in this context, then, is how exactly can digital technologies dis-embed property from these extant peripheral legal, social and material relationships? How can a property relation established and upheld in relational, non-cartographic terms be lifted from its institutional context and fixed into new digitised cartographic and bureaucratic grammars? In this context, we might ask not only how PropTech smoothly subsumes diverse property regimes into capitalistic forms, but also how the political conditions of these property regimes in the majority world and the informational abundance of the periphery might contaminate and contort these projects of digitised enclosure.

Assetisation of peripheral territory

PropTech schemes have their origins in attempts to computerise land records dating back to the late 1980s (Nayak 2013). In 2008 the National Land Record Modernisation Programme (renamed DILRMP in 2016) brought together various land record computerisation programmes. The scheme's long-term aim is to digitise and integrate land records into one central, standardised platform on which information on land ownership and use, encumbrance and geo-referenced plot coordinates are stored and can be automatically mutated and verified. Complementing this central initiative, then, are a host of supplementary state-level endeavours that are primarily engaged in the eradication of customary and communal property systems and their transfer toward private ownership. The central government's Svamitva scheme, introduced at the outset of the paper, is perhaps the most explicit iteration of this ambition. Launched in 2020 by the Ministry of Panchayati Raj (Rural Government) in partnership with the Survey of India and state-level Land Revenue departments, the scheme pledges to map and privately title over six hundred thousand unmapped *abadi* territories across the country, utilising a network of drone, geo-positioning, mapping and platform technologies to distribute ownership rights. *Abadi* lands are the residential territories of rural and urban villages that remain the unmapped, communal property of the village community in much of the country. *Abadi* land is one of a number of examples of the enduring presence of non-privatised, customary land tenure that punctuates landscapes of private real estate in urban and rural India.

The Svamitva scheme relies on the use of drone technology to capture images of possession within the *abadi* boundaries, utilise both human and machine-algorithms to demarcate property on drone images, centralise property data within a publicly accessible online platform, and distribute geo-referenced property titles to those in possession. In doing so the scheme aims to consolidate a new perceptive register in which property is understood as severable, bounded and individually owned. In place of the disaggregated, vernacular and non-cartographic systems of land management, Svamitva's drones, mapping platforms and new continuously operating reference stations consolidate diverse *abadi* tenure into standardised, partible plots that can easily be identified through platform data visualisations and treated as any other fungible financial resource, effecting a form of digitised land assetisation. Indeed, in the act of data transference from paper to the platform, it is figured, the spatiotemporalities of presumptive property systems are eradicated as property relations are cropped and cut to meet normative standards of private property representable in fixed geo-coordinates and standardised alpha-numeric codes.

Once established, the new government platform managing *abadi* records and titles will give investors, financial institutions and state actors full interactive access to *abadi* property, allowing users to layer features such as plot size, land use, encumbrance and ownership details, and initiate property transactions and verifications in real time on the platform, refiguring peripheral sites in the digitised nomenclature of global real estate (Rogers 2016).

Ultimately, the interdigitation of Svamitva-issued property titles into an integrated digital property registry being established under DILRMP will allow larger institutional investors to browse, query and trade property at arm's length, enabling investors to acquire under-valued lands, previously curtailed by their *abadi* status, and consolidate new asset-classes wherein core market functions – from risk-scoring to land valuation – are managed through automated technological processes. The assetisation of peripheral lands has been cited by domestic venture capitalists as a key challenge of the coming decade (Blume Ventures 2025), and despite the incomplete progress of Svamitva, the State Bank of

India has already attributed an uptick in peri-urban mortgage disbursals to the distribution of property cards under the scheme (SBI Research 2022). In addition, there are a host of data analytics firms already utilising data from property digitisation schemes to construct automated land risk and valuation models (Seetharaman 2020). This new commercial stream in public data analytics is made possible by the efforts of government schemes like Svamitva to establish rural assetisation.

There are, in addition, a series of ancillary platform applications that are banking on the asset-classes produced under the scheme. Svamtiva's use of camera and sensor-armed drones to conduct cadastral mapping is producing large-scale territorial data visualisations that can be used for rural governance and infrastructural planning programmes. These visualisations, it is hoped, will strengthen rural state capacity and streamline spatial planning strategies across government departments, while the liberalisation of guidelines on the use of spatial and land data and the creation of a central property registry will enable the private sector to extract public data and build commercial geographic information system-based services (GIS) and applications based on property holdings.

Importantly, processes of assetisation inaugurated under Svamitva aim to replace corrupt, elite-captured governance processes with politically neutral technological systems, transposing local arbitration of the bundle of rights, powers and exclusions inherent to rural property relations to the platform where distant human and non-human actors are tasked with managing the seamless circulation of property and property data as resources through India's public digital infrastructure. This process, for scheme officials, carries moral developmental purpose. The inauguration of digitised private ownership in the *abadi*, it is argued, provides landholders with transparent, legally assured information regarding land ownership, which, in line with the popular neoliberal developmental approach of Hernando de Soto (2000) and others, enables landholders to participate in credit-led modes of socio-economic development (Das 2015; Roy 2012). In this regard, Svamitva aims to eradicate existing institutions of land management and refigure property not only in more clear digital visualisation platforms, but as individuated private assets that can, it is promised by scheme literature, 'be leveraged as financial assets'.

And yet, despite the scheme's claims to celestial techno-neutrality, the fate of digitised assetisation under the scheme – the technological act of transferring existing land-use systems into fixed data categories defined by private ownership – rests upon earthly deliberation and is dependent upon analogue systems of property governance at almost every stage, imbuing assetisation with all manner of trouble (Cowan 2021).

While conducting ethnographic fieldwork in the state of Haryana in 2023 alongside field officials charged with constructing digitised maps under the Svamitva scheme, I found that the consolidation of demarcated plots of possession and distribution of individuated land titles within otherwise unmapped and collectively titled *abadi* territories involved a host of localised deliberation and arbitration that was not solely performed computationally. Drone visuals of the built-up area of an *abadi* territory acted as the primary metric for adjudicating ownership – whatever can be seen (by the drone) must be owned – in ways that forced village residents to make their collective, familial and often non-spatial uses of land visually, aerially clear.

This process of visualisation-as-ownership involves a contested process of deciding 'who owns what' and deliberating over historic possession, disaggregating collective, ancestral and joint-family properties, resolving competing claims and values, and reifying historically developed social inequalities – all by the very customary institutions that digitisation is intended to displace. These acts of deliberation are not only deeply politicised but also profoundly socio-technical, involving work to render customary land uses sensible and actionable as private property (Mitchell 2002).

Let us take the primary operational phases of Svamitva. In the first phase, scheme officials must decide which villages are eligible under the scheme, which villages count as villages, what counts as the boundary of the village and so on. During the second phase, local revenue officials must walk the village with residents, marking out the external boundary of the *abadi* in chalk on the ground – a process referred to as 'chunna-marking'. Chunna-marking is a highly malleable and value-laden exercise involving categorical judgements about how to translate the boundary present on cloth cadastral maps onto the ground, often by approximating parcel shapes – points, cuts, curves; how to adjudicate the formal outer boundary of *abadi* territories that, especially in peri-urban and urban areas, have been informally expanded through an iterative process of village urbanisation (Cowan 2022). These preliminary stages, orchestrated by the public institutions so firmly critiqued as biased by the scheme literature, establish the spatial terms of digitisation processes on the ground.

Within these chalked-out boundaries, village residents are subsequently invited to stage their possession claims to collective village land in strictly spatial and visual terms. Any claims which are historic or common in nature, unsettled or not physically expressed in built-up possession are wiped away and rendered peripheral by the presentist demands of visualising technology used under the scheme. This process, in other words, refigures the rules of property and propertied citizenship in ways which compel residents to align their property claims into data systems. These ground-truth lines are equally undergoing constant drafting and redrafting as they move across different draft stages between state offices, which take opposing positions on how precisely to count possession and territory. In this regard the project of assetisation relies upon the local and vernacular politics of counting – what Martin and Lynch (2009) refer to as 'counting as', where legitimate membership in a domain is dependent upon value-laden judgements and, in our case, visually determined judgements over what counts as property and whose property counts. Of course, this leaves the door open for those very same locally powerful institutions that schemes like Svamitva are designed to displace to use their local clout to establish 'ground-truths' that reflect their own interests. For these reasons, the ambition toward assetisation core to the Svamitva scheme not only has the potential to re-establish already existing propertied and institutional biases of the vernacular land market of the *abadi*, to renovate existing informational peripheries produced through bureaucratic practices of record, but is also simultaneously introducing new metrics of peripheralisation here determined by drone visuality.

Platformisation of peripheral property

At the centre of all current property digitisation projects in India is the platformisation of existing systems of property governance onto a central online interface which will act as the single source of authority on land records nationwide. Platformisation provides a way for the previously discussed peripheralised and regionally specific forms of land use and ownership to be reoriented toward standardised data categories, and for property governance to be streamlined into a 'frictionless' (Sadowski 2020) interface facilitating extended state intervention into formerly opaque and localised property systems (Rao 2019). As previously discussed, PropTech platforms in India are distinct from the market-oriented technologies we tend to associate with platform capitalism (Srnicek 2017), and rather take their structure from the various e-governance initiatives that aimed to solve extant political and social challenges through the technological re-networking of rural and urban state services. The double-edged purpose of the platformisation of property governance enjoins this older technosolutionist purpose by supplanting vernacular land management systems with centralised data platforms, with an additional public and commercial remit wherein the platform itself acts as mediator of new forms of citizenship and customer. Singh and Jackson (2021) argue that these new platform infrastructures enact new kinds of mediated citizenship as they distribute elements of people – in all their diversity – into received, interoperable data representations. In this regard the Indian government's new digital platform infrastructure refigures the state as a decentralised, infrastructural service provider, and a vector toward financialised development.

The government of India's digital public property infrastructure will rest on the ULPIN database – a centralised property record authentication database that draws on ownership and longitude-latitude data produced under a range of national digitisation schemes, with an associated geo-referenced shape file, for each land parcel in India. Modelled on Aadhaar, ULPIN is designed to establish a single authority, record and insurance on property rights, at the same time transferring the legal spatiotemporalities of India's *presumptive* property rights systems to the much desired 'conclusive' property title system. Elsewhere, ULPIN and its associated infrastructure provide a base for new experiments in blockchain-based land management systems that promise (and thus far only promise) the complete detachment of land management operations from corrupt and biased human interference. In 2019, for example, the United Nations Development Programme, working alongside private firm Blockscale Solutions, launched its first pilot blockchain-backed, entirely digital land registration management system in Panchkula, Haryana, declaring that a blockchain-managed land registry can 'ensure that all owners have access to a robust and transparent system for completing transactions' (Williams 2020) and allow people 'to collateralise property, get loans, and invest against them' (Oprunenco and Akmeemana 2018).

ULPIN, like most state digitisation programmes, is being led and designed both by contracted consultancy firms (the 'big four' global management consultancy firms are all involved in various state digitisation programmes) and in heavy consultation with the private banking and tech sectors. This tight integration of private-sector interests in property digitisation schemes is partly to do with the overall ambition of such schemes: to create assets that the private sector might invest in.

Indeed, an explicitly stated ambition of such schemes is to imbricate property data systems into broader digital public infrastructures – the 'India Stack' through which a host of public and commercial services utilise the central Aadhaar identity verification infrastructure to extend access to public services and facilitate financialised property transactions (Dattani 2020). India's digital public infrastructure combines the interests of private finance and domestic software firms with the developmental goals of basic service and citizenship provisions (Samdub 2025). On the one hand, the 'public' in *digital public infrastructure* serves to do the discursive heavy lifting to pose extractive data infrastructure projects in welfarist and thereby perhaps more democratic vernaculars. On the other hand, such a conclusion underplays the shift in meaning and purpose of 'public welfare' in the present conjuncture.

For example, in the Indian state of Haryana, property data produced under the scheme is being utilised primarily by a state public welfare authority, the Haryana Parivar Pehchan Patra Authority (HPPPA) (Haryana Family Identity Card Authority). The department's central platform and database, Parivar Pehchan Patra, has utilised property data to algorithmically determine residents' eligibility for public food rations – a primary welfare provision in India – based on shorthand assumptions about the financial value of newly assetised land. That is, basic citizenship rights are not only being determined algorithmically through the interdigitation of various heterogeneous public datasets, but one's eligibility for such rights is now anchored on one's presumed access to finance. As a HPPPA official explained to me during an interview in 2023, 'their [rural residents'] land is monetised now, it is a monetised asset, so they are able to access loans, we have to take this into account'. It is of course these algorithmic decisions concerning the financial value of property that, as I will discuss in the following section, can produce new forms of peripheralisation through disconnection. As the progeny of Prahalad's (2004) 'fortune at the bottom of the pyramid', the rise of digital public infrastructures operates as a crucial heuristic device for understanding the reconfiguration of public welfare, rather counterintuitively, as the central domain of neoliberal capitalism and authoritarian governance.

Almost all of the Digital India schemes rolled out by the National Informatics Centre are waged at the frontlines of the changing relationship between the state and its populations. I would suggest that we are witnessing an ostensibly new era of digitised governance that renders older 'millennial' financial developmentalism of the 2000s (Roy 2010) onto platform technologies. Such platform developmentalism re-poses state welfare as an opportunity to firmly embed an entrepreneurial logic and democratise financial opportunity, while extending the capacity of the state to flexibly curtail citizenships (Dattani 2020).

And yet again, despite optimism that the new platform infrastructures can eradicate inefficiencies in both governance and property markets, all these experiments in property digitisation are reliant on the accuracy of incoming property data. Under the existing system (with some differences from state to state), property ownership and rights are *presumptive*, premised upon one's ability to retroactively prove a successive chain of possession through documentation relating to transfer, tax, inheritance and so on. And as a consequence, property rights are spatially disaggregated, manifest in fragmented sets of government ledgers and certifications, and temporally retroactive, assembled and performed after the fact. The heterogeneity of property's documentary life causes all kinds of problems for initiatives like Svamitva, and ancillary platforms like the HPPPA previously described, which rely on simple, crisp alignment of datasets. Importantly for us, this system embeds knowledge of localised institutions and legal authority in ledgers and paperwork that are not so easily displaced and, as a consequence, layers new platform infrastructure on a lively and malleable set of local governance relations (Cowan 2021). Platformisation, in this context, represents a rather speculative *promise to* bring the retroactive and disaggregated spatiotemporalties of peripheral property into line with a fixed logic of private ownership. In this regard, current state-level digitised land registries remain tied to analogue processes of deed and ground verification (Cowan 2022).

The platformisation of peripheral property is thus under constant interrogation from the diverse institutional relations interred within its operational logic, threatening the veracity of infrastructures like ULPIN to operate as seamless platform mediators of property markets and property governance. For Singh (2019) and Rao (2019), 'the platform state' signals the completion of a process of divestment from postcolonial dirigisme that began in the early 1990s, and instead reorients governance to the task of linking eligible citizens to public services, sorting citizens' eligibility for public services through conditions of data legibility (Singh and Jackson 2021), and utilising real-time data to identify developmental gaps that might be remedied by entrepreneurial, technical and often finance-oriented solutions (Rao 2019; Sarkar 2014). And yet the alignment of provisional and presumptive property forms remains a tricky proposition, requiring a constant return to the local, embedded institutions that digitisation was intended to displace. As such we end up with 'recombinant' (Balakrishnan 2019) state structures composed across the cloud and the ground.

Peripheralisation

As the discussion of the mason at the start of this chapter highlights, the integration of property data produced under PropTech schemes like Svamitva into digital public infrastructures transforms existing state–citizen relations, enabling the state to speculatively assess a citizen's eligibility or access to public services based on clusters of data analytics.

If in the past property rights claims were, as previously discussed, determined by one's ability to navigate the paper politics of local bureaucracies (Hull 2012), under PropTech infrastructures, rights to inhabit, use, inherit and own land are determined algorithmically, premised upon one's ability to align one's property claims with the visual and severable attribute data categories of programmes like ULPIN, DLRMP and Svamitva. As Hasan writes, any and all 'excess in the form of hacks, leaks, and errors in data produce new obligations on individuals needing this data to exercise their basic rights', noting further that a focus on data excess 'offers a vantage point from where to view citizenship practiced and experienced as a tryst with broken data' (Hasan 2022, 122).

One clear moment of excess in property digitisation processes arrives in the desired path-dependent interoperability and co-dependence of the various databases that constitute the digital property infrastructure. From the vantage point of the landholder or user, one's property may be digitised and linked to a ULPIN digital ID only when legal disputes over ownership and location of land have been resolved, when one's multiple share-based ownership has been cleanly parted into drone-identifiable plots, when inconsistencies in Svamitva boundary-marking have been resolved, when one's textual record of ownership is made interoperable with a GIS demarcation of the land plot, and so on. There are, in other words, a host of categorical labours and processes of data transference that must align in order for one to receive an identity on the ULPIN platform. As is perhaps telling from the slow progress made by the government's DILRMP scheme, which has spent over a decade attempting to digitise and consolidate textual and spatial property records across the country, the diversity of property rights and claims is not easily and cleanly translatable to digital registers.

An error or inconsistency in one's property record may lead to the withdrawal of state welfare or services, or foreclose access to vital private services and markets. What's more, when one's property relations do not neatly fit the strictured epistemology of the digital category, one is not only dispossessed of one's property claim, but, under integrated digital infrastructures, also marginalised from a whole host of basic services, undermining the integrity of one's citizenship.

Such peripheralisation is produced in two key ways. The first is the buttressing of existing inequalities in property regimes by digital technologies. For example, as there is no mechanism for incorporating territories adjacent to *abadis* (typically occupied by Scheduled Caste communities) into Svamitva titling schemes, Svamitva stands to re-establish new forms of digitised caste-based property regimes. In addition, there remains no formal mechanism for incorporating collectivised or familial property uses within Svamitva data categories. Furthermore, given women's historic exclusion from property registers in many parts of the country (Baviskar and Mathew 2009), the formalisation of property rights in a centralised platform stands to formalise extant gender inequalities.

Second, peripheralisation occurs due to the variegated form in which citizenship is made legible under data systems. In a discussion of citizenship under India's Aadhaar identification system, Singh and Jackson note that digitised citizenship operates through grades of data resolution, wherein 'certain combinations of data [provide] more detailed and comprehensive data representations of [some] people than others' (Singh and Jackson 2021, 7). The effect of these data combinations produces citizenships-in-resolution, wherein the inability of someone to align their life with data categories renders them 'low resolution', pixellated and less visible to state infrastructures, while a high presence of aligned data renders others 'high resolution' and with greater access to basic citizenship services (Singh and Jackson 2021, 7). Here, digitalised identification schemes produce new rounds of peripheralisation, as discussed across this collection.

There are clear analogies here with how standard protocols for transferring property data under schemes like Svamitva produce differing levels of detail within digital property infrastructures. For example, the fact that a share-held common-use plot cannot be fully represented within Svamitva data categories necessarily renders certain non-individuated claims in 'low resolution' and thus disconnects some from their property uses. Citizenships-in-resolution also point to the ways in which digital governance infrastructures are restructuring the act of citizenship itself, wherein rights are not presumed but must be actively proven through one's legibility in data systems that are, as discussed, designed with a very narrow and particular form of person (asset-holder) and property (private asset) in mind. When data transference is interrupted, contains errors, duplicates or glitches, citizens are forced to actively repair their data, determining their access to substantive propertied citizenship (Hasan 2022).

While property digitisation and platformisation remain in their early stages, we can already see the various permutations of residents' property relations appearing in 'high' and 'low resolution' in state infrastructures, albeit with differing consequences. While appearing in 'high resolution' may open them up to the benefits and risks of financial markets, it may also subject them to undue state calculation and surveillance. Appearing in low resolution, on the other hand, may skirt such surveillance, but – as in the case of the mason – exposes residents to new digitalised forms of dispossession and peripheralisation.

As discussed above, the transference of property rights to the digital registry may equally expose people to unwanted surveillance and oversight, which may also lead to the removal of erstwhile privileges and rights, legitimately held or otherwise. As Singh and Jackson note, people often choose illegibility from the state, and what we might think of as data abjection 'in the face of long histories of surveillance and tracking' (Singh and Jackson 2021, 8) might be equally understood as tactics to evade state oversight.

Digitising peripheral property

This chapter has explored how efforts to digitise and assetise property in contemporary India are contributing to the production of new forms of informational connection and peripheralisation. The chapter has provided an overview of the character and deployment of property technologies in contemporary India, examining how such technologies in this context are explicitly being put to work to refashion peripheral property and property regimes that stand in the way of both financialised real-estate markets and platform government in much of the country. In doing so, the chapter has aimed to (in part) provincialise current understandings of PropTech to their North Atlantic origins and explore the character of property technologies in contexts in which private property is far from the norm.

Following the themes of this issue, we might think of state-led PropTech initiatives as inaugurating new forms of informational centralisation and peripheralisation that are rewriting the terms of propertied citizenship in the present conjuncture. Here, the digitisation, assetisation and platformisation of rural and peri-urban property, otherwise communal in ownership and institutionally plural, engage in a process of informational centralisation and informational peripheralisation. On the one hand, the central government's property digitisation programmes aim to centralise: to render previously unmapped and unmappable, uncounted and uncountable property relations, mappable and countable to the registers of both the state and financial institutions. The reorganisation of customary and communal property regimes is justified as a distribution of ownership rights, and a vehicle to create new forms of propertied citizenship for landed members of rural communities. However, in order to achieve this, state and private-sector actors must engage in acts of data transference that eradicate the legal authority of non-privatised property relations – all manner of relations to land and territory that do not align with core data categories, in turn producing new forms digital marginality.

In examining the attempted capture and reconfiguration of peripheral property, the chapter has identified three core features. The first is assetisation – the explicitly stated desire of property technology schemes, operating under the guise of 'clarity' and techno-objectivity, to assetise residual non-normative property regimes across the country. The second is platformisation – the desire to reconfigure vernacular property management systems onto interoperable software interfaces that facilitate digitised territorialisations of state authority and new avenues of data commercialisation. Finally, the chapter explored how these first two conditions have reposed citizen–state relations, producing (dis)connected forms of digitised propertied citizenship that appear in graduated 'resolution' within state infrastructures. At each juncture the chapter has highlighted how the ambitious, techno-neutral imaginaries of PropTech in India are undermined by a constant requirement to renovate existing vernacular property systems and propertied inequalities, infusing property digitisation endeavours with a degree of trouble, contamination and rewiring that leaves the fate of PropTech endeavours open to political resettlement.

References

Amoore, Louise. 2020. *Cloud Ethics: Algorithms and the attributes of ourselves and others*. Durham, NC: Duke University Press.

Balakrishnan, Sai. 2019. Recombinant urbanization: Agrarian–urban landed property and uneven development in India. *International Journal of Urban and Regional Research* 43(4): 617–32. https://doi.org/10.1111/1468-2427.12790.

Bathla, Nitin. 2022. Extended urbanisation and the politics of uncertainty: The contested pathways of highway corridors in India. *The Geographical Journal* 190(1). https://doi.org/10.1111/geoj.12441.

Baviskar, Baburao Shravan and George Mathew. 2009. *Inclusion and Exclusion in Local Governance: Field studies from rural India*. Thousand Oaks, CA: Sage Publications.

Benjamin, Solomon. 2005. Touts, pirates and ghosts. *Sarai Reader* 5: 242–54. Accessed February 2025. https://www.yumpu.com/en/document/read/19888014/touts-pirates-and-ghosts-sarai.

Benjamin, Solomon and Bhuvaneswari Raman. 2011. Illegible claims, legal titles, and the worlding of Bangalore. *Revue tiers monde* 2: 37–54. Accessed February 2025. https://shs.cairn.info/revue-tiers-monde-2011-2-page-37?lang=en&tab=texte-integral.

Bhan, Gautam. 2019. Notes on a southern urban practice. *Environment and Urbanization* 31(2): 639–54. https://doi.org/10.1177/0956247818815792.

Blomley, Nicholas. 2005. Law, property, and the geography of violence: The frontier, the survey, and the grid. *Annals of the Association of American Geographers* 93(1): 121–41. https://doi.org/10.1111/1467-8306.93109.

Blume Ventures 2025. *Indus Valley Annual Report 2025.* Accessed February 2025. https://blume.vc/reports/indus-valley-annual-report-2025.

Bowker, Geoffrey C. and Susan Leigh Star. 2000. *Sorting Things Out: Classification and its consequences*. Cambridge, MA: MIT Press.

Buzan, Barry and George Lawson. 2013. The global transformation: The nineteenth century and the making of modern international relations. *International Studies Quarterly* 57(3): 620–34. https://doi.org/10.1111/isqu.12011.

Chatterjee, Partha. 2004. *The Politics of the Governed: Reflections on popular politics in most of the world*. New York: Columbia University Press.

Cowan, Thomas. 2021. Uncertain grounds: Cartographic negotiation and digitized property on the urban frontier. *International Journal of Urban and Regional Research* 45(3): 442–57. https://doi.org/10.1111/1468-2427.13016.

Cowan, Thomas. 2022. *Subaltern Frontiers: Agrarian city-making in Gurgaon*. Cambridge: Cambridge University Press.

Das, Raju J. 2015. Critical observations on neo-liberalism and India's new economic policy. *Journal of Contemporary Asia* 45(4): 715–26. https://doi.org/10.1080/00472336.2014.1003143.

Dattani, Kavita. 2020. 'Governtrepreneurism' for good governance: The case of Aadhaar and the India stack. *Area* 52(2): 411–19. https://doi.org/10.1111/area.12579.

de Soto, Hernando. 2000. *The Mystery of Capital: Why capitalism triumphs in the West and fails everywhere else*. New York: Basic Books.

Faxon, Hilary Oliva and Courtney T. Wittekind. 2023. Livestreamed land: Scams and certainty in Myanmar's digital land market. *Environment and Planning D: Society and space* 42(4): 512–33. https://doi.org/10.1177/02637758231205958.

Fields, Desiree. 2022. Automated landlord: Digital technologies and post-crisis financial accumulation. *Environment and Planning A: Economy and space* 54(1): 160–81. https://doi.org/10.1177/0308518X19846514.

Ghertner, D. Asher. 2014. India's urban revolution: Geographies of displacement beyond gentrification. *Environment and Planning A: Economy and space* 46(7): 1554–71. https://doi.org/10.1068/a46288.

Ghertner, D. Asher. 2015. Why gentrification theory fails in 'much of the world'. *City* 19(4): 552–63. https://doi.org/10.1080/13604813.2015.1051745.

Goodfellow, Tom, 2020. Political informality: Deals, trust networks, and the negotiation of value in the urban realm. *Journal of Development Studies* 56(2): 278–94. https://doi.org/10.1080/00220388.2019.1577385.

Goswami, Amlanjyoti, Deepika Jha, Kaye Lushington, Sahil Sasidharan and Sudeshna Mitra. 2017. *Land Records Modernisation in India: An institutional, legal & policy review.* Bengaluru: Indian Institute for Human Settlements.

Government of India. 2020. PM's address at the launch of physical distribution of property cards under the SVAMITVA scheme. Accessed February 2025. https://pib.gov.in/PressReleseDetailm.aspx?PRID=1663540&reg=3&lang=1.

Hall, Derek, Philip Hirsch and Tania Murray Li. 2011. Introduction. In *Powers of Exclusion: Land dilemmas in Southeast Asia*, 1 26. Singapore: National University of Singapore Press.

Harriss-White, Barbara. 2003. *India Working: Essays on society and economy*. Cambridge: Cambridge University Press.

Hasan, Nafis. 2022. 'Slow violence' and vacant citizenship: The excesses of India's digital governance. In *Overload, Creep, Excess: An internet from India*, edited by N. Shah, A. Rajadhyaskha and N. Hasan, 121–62. Amsterdam: Institute of Network Cultures.

Hasan, Nafis. 2024. Citizen labor: Correcting data and creating value in an Indian land records database. *American Ethnologist* 51(3): 376–87. https://doi.org/10.1111/amet.13303.

The Hindu. 2020. SVAMITVA scheme will unleash productive forces in agriculture and non-farm rural activities. 20 October. Accessed February 2025. https://www.thehindubusinessline.com/opinion/editorial/svamitva-scheme-will-unleash-productive-forces-in-agriculture-and-non-farm-rural-activities/article32902689.ece.

Hsing, You-tien. 2010. *The Great Urban Transformation: Politics of land and property in China*. Oxford: Oxford University Press.

Hull, Matthew. 2012. *Government of Paper: The materiality of bureaucracy in urban Pakistan*. Berkeley, CA: University of California Press.

Jefferson, Brian Jordan. 2018. Predictable policing: Predictive crime mapping and geographies of policing and race. *Annals of the American Association of Geographers* 108(1): 1–16. https://doi.org/10.1080/24694452.2017.1293500.

Jonnalagadda, Indivar and Thomas Cowan. 2024. City drafting: Property-making and bureaucratic urbanism in South Asia. *City* 28(1–2): 7–23. https://doi.org/10.1080/13604813.2024.2321024.

Jonnalagadda, Indivar, Ryan Stock and Karan Misquitta. 2021. Titling as a contested process: Conditional land rights and subaltern citizenship in South India. *International Journal of Urban and Regional Research* 45(3): 458–76. https://doi.org/10.1111/1468-2427.13002.

Kanai, J. Miguel and Seth Schindler. 2022. Infrastructure-led development and the peri-urban question: Furthering crossover comparisons. *Urban Studies* 59(8): 1597–617. https://doi.org/10.1177/00420980211064158.

Kshetri, Nir. 2022. Blockchain as a tool to facilitate property rights protection in the Global South: Lessons from India's Andhra Pradesh state. *Third World Quarterly* 43(2): 371–92. https://doi.org/10.1080/01436597.2021.2013116.

Kumar Sharma, Shiv. 2020. Modi interacts with Yamunanagar's mason during distribution of property cards. *Tribune India*, 11 October. Accessed February 2025. https://www.tribuneindia.com/news/haryana/modi-interacts-with-yamunanagars-mason-during-distribution-of-property-cards-154235.

Martin, Aryn and Michael Lynch. 2009. Counting things and people: The practices and politics of counting. *Social Problems* 56(2): 243–66. https://doi.org/10.1525/sp.2009.56.2.243.

McElroy, Erin and Manon Vergerio. 2022. Automating gentrification: Landlord technologies and housing justice organizing in New York City homes. *Environment and Planning D: Society and space* 40(4): 607–26. https://doi.org/10.1177/02637758221088868.

Migozzi, Julien. 2020. Selecting spaces, classifying people: The financialization of housing in the South African city. *Housing Policy Debate* 30(4): 640–60. https://doi.org/10.1080/10511482.2019.1684335.

Migozzi, Julien, Michael Urban and Dariusz Wójcik. 2023. 'You should do what India does': FinTech ecosystems in India reshaping the geography of finance. *Geoforum* 151. https://doi.org/10.1016/j.geoforum.2023.103720.

Mitchell, Timothy. 2002. *Rule of Experts: Egypt, techno-politics, modernity*. Berkeley, CA: University of California Press.

Nayak, Pradeep. 2013. Policy shifts in land records management. *Economic & Political Weekly* 48(24). Accessed February 2025. https://www.epw.in/journal/2013/24/notes/policy-shifts-land-records-management.html.

Oprunenco, Alexandru and Chami Akmeemana. 2018. Using Blockchain to make land registry more reliable in India (blog). *LSE Business Review*, 13 April. Accessed February 2025. https://blogs.lse.ac.uk/businessreview/2018/04/13/using-blockchain-to-make-land-registry-more-reliable-in-india.

Phan, Thao and Scott Wark. 2021. Racial formations as data formations. *Big Data & Society* 8(2). https://doi.org/10.1177/20539517211046377.

Prahalad, C. K. 2004. *The Fortune at the Bottom of the Pyramid*. Upper Saddle River, NJ: Prentice Hall.

Rao, Ursula. 2019. Response to 'The Aadhaar debate: Where are the sociologists?'. *Contributions to Indian Sociology* 53(3): 431–40. https://doi.org/10.1177/0069966719861759.

Richter, Christine. 2011. In-tensions to infrastructure: Developing digital property databases in urban Karnataka, India. *Environment and Urbanization Asia* 2(2): 205–22. https://doi.org/10.1177/097542531100200205.

Rogers, Dallas. 2016. Uploading real estate: Home as a digital, global commodity. In *Housing and Home Unbound: Intersections in economics, environment and politics in Australia*, edited by Nicole Cook, Aidan Davison and Louise Crabtree, 23–38. Abingdon: Routledge.

Roy, Ananya. 2010. *Poverty Capital: Microfinance and the making of development*. Abingdon: Routledge.
Roy, Ananya. 2012. Ethnographic circulations: Space–time relations in the worlds of poverty management. *Environment and Planning A: Economy and space* 44(1): 31–41. https://doi.org/10.1068/a44180.
Sadowski, Jathan. 2020. The internet of landlords: Digital platforms and new mechanisms of rentier capitalism. *Antipode* 52(2): 562–80. https://doi.org/10.1111/anti.12595.
Safransky, Sara. 2020. Geographies of algorithmic violence: Redlining the smart city. *International Journal of Urban and Regional Research* 44(2): 200–18. https://doi.org/10.1111/1468-2427.12833.
Samdub, Mila. 2025. 'Digital public infrastructure' at a turning point: From definitions to motivations. Open Future Foundation. Accessed February 2025. https://openfuture.eu/publication/digital-public-infrastructure-at-a-turning-point.
Sarkar, Swagato. 2014. The unique identity (UID) project, biometrics and re-imagining governance in India. *Oxford Development Studies* 42(4): 516–33. https://doi.org/10.1080/13600818.2014.924493.
SBI Research. 2022. *Special Report on Emerging Trends in Residential Housing*. Accessed February 2025. https://sbi.co.in/documents/13958/25272736/Special+Report+on+Emerging+Trends+in+Residential+Housing+%283%29.pdf.
Seetharaman, G. 2020. Whose land is it anyway? *The Ken*, 18 August. Accessed February 2025. https://the-ken.com/story/whose-land-is-it-anyway.
Sikor, Thomas and Christian Lund. 2009. Access and property: A question of power and authority. *Development and Change* 40(1): 1–22. https://doi.org/10.1111/j.1467-7660.2009.01503.x.
Singh, Ranjit. 2019. Give me a database and I will raise the nation-state. *South Asia: Journal of South Asian Studies* 42(3): 501–18. https://doi.org/10.1080/00856401.2019.1602810.
Singh, Ranjit and Steven Jackson. 2021. Seeing like an infrastructure: Low-resolution citizens and the Aadhaar Identification Project. *Proceedings of the ACM on Human-Computer Interaction* 5 (CSCW2): 1–26. https://doi.org/10.1145/3476056.
Siwach, Sukhbir. 2023. Parivar Pehchan Patra: Over 8 lakh BPL families lose ration cards. *The Indian Express*, 22 February. Accessed February 2025. https://indianexpress.com/article/cities/chandigarh/parivar-pehchan-patra-over-8-lakh-bpl-families-lose-ration-cards-8459384.
Srnicek, Nick. 2017. *Platform Capitalism*. Chichester: John Wiley & Sons.
Verdery, Katherine. 1999. Fuzzy property: Rights, power and identity in Transylvania's decollectivization. In *Uncertain Transition: Ethnographies of change in the postsocialist world*, edited by M. Burrawoy and K. Verdery, 53–81. Lanham, MD: Rowman & Littlefield.
Williams, Anthony. 2020. India's Land Registry on blockchain. Blockchain Research Institute, 18 December. Accessed February 2025. https://www.blockchainresearchinstitute.org/project/indias-land-registry-on-blockchain.

Theme III

Informational agency from the peripheries

10

Paper truths, digital authentication: India's interstate migrants and the path to urban citizenship

Shoshana Goldstein

India's interstate migrants straddle informational peripheries, living between rural villages and the literal and figurative margins of the cities where they work to diversify their incomes (Kantor 2018; Breman 1996). This dual existence depends heavily on paper documentation to establish residency and citizenship. However, this reliance on documentation limits access to urban benefits and often serves as a tool of structural violence against India's migrant poor (Abbas 2016; Gupta 2012; Hull 2012; Sriraman 2018). This chapter explores the urban informational periphery of Delhi and Gurugram, where migrants navigate everyday needs and bureaucratic demands, facing institutionalised barriers to accessing critical information and services.

For decades, civil society groups have worked to lessen these barriers of distance as well as connectivity, as Datta and Hoefsloot set out in the introduction to this volume. These groups assist households to obtain 'proof documents' and the formalisation of their residency while offering workshops to develop skills that can travel with migrants. Such programmes mirror government initiatives aimed at developing techno-literacies in the Indian labour force more broadly (Gooptu 2019; Upadhya 2016), and reflect the hope that digital record-keeping might standardise and validate documents, reducing bureaucratic disputes. However, digital platforms can replicate and even exacerbate barriers and do not necessarily relieve the burdens of establishing 'paper truths' for migrants or the urban poor.

The term 'paper truths', coined by Tarlo, encapsulates the critical disconnect between official documents and reality, suggesting that 'their relationship to reality is open to manipulation and fraught with ambiguity' (Tarlo 2001, 69). This chapter argues that the paper truths phenomenon does not dissolve with the transition to a digital medium but evolves to produce similar disconnects between the actual and the official. Rather than increasing transparency, the digital realms of information gathering and sharing may compound the opacity and delays of paper truths. Using digitally authenticated tools such as Aadhaar or the 'One Nation One Ration Card' scheme to access the Public Distribution System (PDS) also sidesteps the question of establishing local citizenship, favouring portability over enabling migrants to *become* locals. Compounded by privacy issues and the mission creep of citizen surveillance, such schemes therefore present the downsides of being accounted for by the state (Dixon 2017). It remains to be seen whether these modes of digital authentication, which do not address the root economic causes of migration, lead to greater inclusivity or ease the pressure for rural–urban migration strategies.

The informational periphery in this chapter encompasses the Delhi Region, focusing on Kapashera colony, which lies between Delhi's National Capital Territory and Gurugram, Haryana. Kapashera is peripheral both literally and relationally, forming an adjoining boundary between two cities and two states for people inhabiting many peripheries at once. Through ethnographies and interviews with migrants in 2023, and digital ethnographies conducted via WhatsApp during the COVID-19 pandemic, this chapter reveals the contradictions of digital authentication. The potential for digital tools to save money and time is often offset by the high costs of digital access itself. This not only diminishes the economic benefits of digital tools but also fails to reduce the significant time migrants must spend away from wage-earning work when navigating bureaucratic processes. Moreover, gaps in digital literacy and access reinforce existing inequities, undermining the inclusive potential of digitalisation.

During India's pandemic lockdowns, urban local bodies were urged to expedite the digitisation of migrant residency (*Hindustan Times* 2020; *Economic & Political Weekly Engage* 2020). While the pandemic accelerated digital technology use, its impact on migrants' access to government services remains unclear. The informational periphery instead became a space for novel survival strategies and hierarchies of digital literacy and entrepreneurship. This chapter reimagines the periphery from below, highlighting how migrants navigate digital environments amid top-down digitalisation efforts. Age and gender significantly influence engagement with both digital platforms and paper-based systems. Notably, female married migrants in the study, often with limited education and discouraged from developing digital skills, faced greater challenges in managing digital records and online forms, deepening marginalisation in the informational periphery.

From paper truths to digital authentication

Tarlo's (2001; 2003) notion of 'paper truths' provides a shorthand for understanding how the bureaucratic state values documentation over other means of validating information. Deeds, sterilisation consent forms and other records as paper truths suggest the existence of many truths, including a contrasting lived reality for individuals who may have signed these documents. This system of paper truths emerges from a tradition of knowledge production in colonial and postcolonial public administration services, revolving around the census, the original ration card schemes, and various other technologies of social accounting (Routray 2022; Sriraman 2018; Raman 2012; Dirks 2011; Stoler 2008; Bayly 2000; Anderson 2006; Chatterjee 1993; Cohn 1987).

Similarly, Raman (2012) coins the term 'paper reality' to describe South Indian scribe culture. Within the so-called 'document raj', the cultural significance of the scribe translated into the daily grind of the imperial bureaucrat. In this context, documentation (specifically paper-based) gains legitimacy from the state and citizens who conform to its demands, often at the expense of actual events or facts. Paperwork transcends mere documentation; it becomes a tool of power and control, often disregarding other forms of evidence and reality.

This bureaucratic system, characterised by its fixation on documentation, and the practices surrounding paper truths, often perpetuates and exacerbates the vulnerabilities of India's poor. While the state is supposed to represent and care for the poor, how local bureaucrats carry out poor-serving policies can perpetrate obscured and seemingly incidental violence. Such violence, at first glance, may not appear to be the result of deliberate state neglect, maladministration or failure to govern, shielding administrators from culpability (Gupta 2012). Consequently, this type of incidental violence manifests in the formation of an informational periphery – a figurative space where marginalised populations encounter both the analogue and digital demands of a bureaucratic state (Datta 2024).

While the shift from paper-based to born-digital identification methods is significant, the bureaucratic fixation on the accuracy and integrity of information persists across media, whether through signatures, e-signatures or biometric data (such as fingerprints). Despite evolving methods of identification, Tarlo's (2003) work demonstrates how pre-digital documents produced informational peripheries in Delhi during the emergency period (June 1975–March 1977). Paper truths remind us of the historical continuities between the analogue and digital realms of red tape.

The disparity between what the documentation portrays and actual events often leads to injustices for those on the informational periphery. These injustices form part of a broader pattern in South Asian and European colonial governance agencies, such as the Indian Civil Service established under British Crown rule, which formed the basis for the Indian Administrative Service after independence. These systemic issues reveal how the bureaucracy, through its processes of documentation and authentication, creates and maintains informational peripheries, perpetuating a cycle of vulnerability and exclusion for the poor.

Authentication: digital hopes and realities

'Authentication' refers to a set of practices and procedures that confirm or deny the validity of identities. It involves the literal process of digitisation – specifically, the interlocking processes of enrolment and database 'seeding' (Singh 2019). The enrolment process collects traditional paper documentation and biometric data, while seeding converts this analogue information into digital formats. There is no fundamental departure from paper truths in this process, but rather a recapitulation of them. As Sriraman observes in her ethnography of Aadhaar enrolment, 'paper-based ID documents did not disappear after their information was transcribed for generating the unique number. If anything, the process of document verification intensified' (Sriraman 2018, 264). Verifying identities within public service databases and the platforms of a digitalised state is particularly crucial for what Singh and Jackson (2021) describes as 'low resolution citizens'. These individuals often face barriers due to poor data quality, changes in biometric data over time or systemic failures in accurately reading fingerprints (Henne 2019).

These difficulties in demonstrating authenticity raise questions about the nature of identity as an object of knowledge production and a digital doppelganger, subject to the information architectures within government record keeping. India grapples with both distinct technical challenges and enduring cultural narratives about record falsification (Srivastava 2012). In addition to falsification, the duplication of records poses the possibility of double-dipping within the PDS. Thus, a general expectation of duplicate records creates an air of suspicion around the enrolment and seeding processes developed to 'deduplicate' such records. For instance, the Aadhaar system and the 'One Nation One Ration Card' initiative illustrate efforts to streamline identity verification by creating a 'root identity' around which all other forms of identification are linked (Singh 2019). The latter leverages the Aadhaar biometric system to authenticate beneficiaries through an Integrated Management of the PDS system to ensure national operability of the PDS and intra-state portability of ration cards.[1] Migrant households would thus ideally be able to access the food to which they are entitled from any affordable shop located anywhere in the country.

However, errors in developing 'born-digital' information and the seeding process can be difficult to undo. Incorrect information, such as misspelt names or minor typos, has inked errors into the record that are difficult to correct without inadvertently admitting to fraud (Srivastava 2012). Minor errors in authentication can affect tens of millions of users, emphasising that authentication is not merely technical but a socio-political act that embodies new iterations of paper truths (Singh 2019). The errors that arise in the process of seeding digital information due to crashed servers, weak connectivity in rural areas and unrecognisable fingerprints (partly due to the corrosive nature of migrant occupations) continue this tradition.

Ultimately, assumptions about the efficiency of digital infrastructures belie the paper truths that undergird the entire process. The problem is not one of digital versus paper, but rather that rigid methods of verification cannot accommodate or correct human error, and therefore often must double down on it. As one participant explained, when their name was misspelt on a form, they decided it was easier to embrace the error as an intentional feature of their signature going forward. Such errors eventually lead to exclusion from state benefits, perpetuating the structural violence of bureaucracies as they dismiss potential beneficiaries with the complexities of red tape, in both paper and digital formats.

The contradictions of digital authentication

The digitalisation of governance poses significant challenges for India's 'low resolution' citizens, including interstate migrants. Digital platforms controlling access to geographic and civic data marginalise individuals who are not fully integrated into these systems. The widespread use of these platforms in urban life intensifies conflicts over who is included or excluded, who has control, and who maintains sovereignty over personal data and an 'informational right to the city' (Shaw and Graham 2017).

As digital platforms shape everyday economic and social transactions, exerting control over their users (Poell et al. 2019), lack of documentation restricts the use of government programmes such as the PDS, making it difficult for migrants to secure essential food items and financial assistance during emergencies (Agarwal and Agnihotri 2022). Without proper documentation, migrants face barriers in accessing healthcare (Santalahti et al. 2020). Migrant children encounter obstacles to enrolling in school and accessing education (Naik et al. 2022).

In terms of employment, informal migrant labourers have minimal job security and low wages. The lack of formal employment exacerbates their susceptibility to exploitative labour practices, including long hours and hazardous conditions, with limited legal protections and fears of retaliation (Datta (Amrita) 2020; Hirudayaraj et al. 2024; Jayaram and Varma 2020). Some labourers may pay bribes or fees to navigate bureaucratic hurdles or secure documentation, which adds to their financial burden. Securing housing or rental accommodation through oral agreements does not generate a paper trail, leaving migrants living in informal settlements under inadequate living conditions without written lease agreements to use as 'proof documents' for verification of their identity and claims (Naik 2019). Without formal residency status, migrants cannot seek representation in civic processes, including local elections, contributing to their marginalisation as political outsiders in their host states (Abbas 2016).

Non-governmental organisations (NGOs) and community-based organisations (CBOs), including trade unions, play a role by facilitating access to tools and platforms for accessing public services, jobs, education and housing. They assist migrants in obtaining documentation and understanding their rights within the digital sphere, acting as intermediaries between the state and the individual (De Wit and Berner 2009; Medappa 2023). Informal networks, patronage relationships and mutual aid also play important roles here, using tools such as WhatsApp for communication (Maddox and Kanthawala 2023).

Despite the potential for such tools to enhance access to government benefits, according to the Director of Agrasar, a migrant-serving NGO, more than 90 per cent of migrants in the study site do not register locally for benefits, primarily due to the aforementioned documentation issues and a lack of knowledge about the processes involved (Interview, 4 August 2023). The extensive use of mobile phones and other digital technologies contrasts sharply with these figures, suggesting a disconnect between everyday digital literacies and the pervasiveness of mobile phones (Mukhopadhyay et al. 2023), and the documentation challenges involved in using government platforms.

The challenges of authentication and verification in digital systems reveal a troubling continuity with the past. While digitisation aims to streamline access to public services, authentication processes often replicate and exacerbate the exclusionary practices of the paper-based era. Reliance on biometric data and inflexible verification methods have left many interstate migrants entangled in costly processes of mitigating technical errors, while dealing with administrative neglect (Ray 2024).

Informational peripheries in Gurugram and Kapashera

This research employs an ethnographic approach to understand the informational periphery for interstate migrants, focusing on their experiences navigating India's digital bureaucracies and attempts to authenticate their identities, or decisions to abandon the process altogether. The research draws from interviews with a cohort of 25 migrant workers living in the Kapashera colony on the Delhi–Gurugram border during the summer of 2023.

As a researcher, I relied on a network of local informants and connections to migrant-serving NGOs that manage programmes addressing the needs, skilling and residency status issues of interstate migrants. However, my position as an American academic without ties to the local government, familial or caste affiliation in India significantly influences the research process. This status grants me a unique perspective but also presents challenges in establishing research relationships and trust with participants. My presence in urban slums is often viewed with curiosity or suspicion. For this research, I collaborated closely with a research assistant – a former migrant in the garment sector and a member of a lower-caste community from rural Uttar Pradesh. His embeddedness in the research context, proficiency in Hindi and English and survey experience with migrant-serving NGOs were invaluable. Together, we crafted an interview protocol, identifying key participant demographics to ensure a representative cohort covering diverse migrant experiences. We liaised with several NGOs to identify potential participants. We recruited through snowball sampling within the garment factory labour community. The regional composition of participants was primarily Bihari, with some from West Bengal and Uttar Pradesh, reflecting the typical demographic breakdown of the enclave's textile manufacturing workforce.

Given heightened animus against Muslims in the Delhi National Capital Region in the summer of 2023, research was conducted in neutral third spaces to ensure the safety and privacy of respondents. Interviews were conducted from the offices of several NGOs providing instruction on digital literacy to adults and local youth. This approach was vital for protecting participants from potential backlash from landlords, neighbours or local authorities. Interviews with migrants were recorded in Hindi and transcribed into English. The participants ranged in age from 19 to 56. A quarter of participants were female, typically married with children. Interviews were conducted in the evenings and on weekends to avoid conflicts with household responsibilities and working hours. Interviews began with demographic questions to establish household composition, living standards and migration strategies. This was followed by open-ended discussions centred around social histories, usage of digital technologies and experiences with paperwork and accessing the PDS.

Buying urban citizenship: the high costs of paper truths

On a humid monsoon evening, I met the first interviewees in a bubble-gum-pink NGO digital classroom, its walls covered with inspirational quotes from tech moguls like Jeff Bezos and Elon Musk (Figure 10.1). The absence of a Mark Zuckerberg quote on the wall belied his influence via Facebook, and most importantly, WhatsApp. The use of WhatsApp was supplemented by YouTube, a source for news, and Flipkart, an online shopping app. The posters reflected the ubiquity of digital platforms in the lives of migrants, underscoring the critical role of technology in both enabling and complicating access to services and information.

We chose this classroom to avoid landlords and neighbours likely to retaliate. This neutral space facilitated discussions over chai or lemon soda, offering privacy amid regional tensions affecting Muslims and Bengali-speaking migrants. Hindu migrants from Uttar Pradesh and Bihar spoke with varied Hindi proficiency. One interviewee quoted a local saying that had to be translated to English before I grasped it: 'Every two miles the water doth change, and every four the dialect.'

Figure 10.1 NGO digital literacy classroom, India. Photograph: Shoshana Goldstein, 2023.

Respondents worked mostly in the local garment factories, with some engaged in home-based tailoring gigs. Interviews always began with questions about the basics and a survey of household expenses. Of course, transportation and proximity to work were vital. Water had to be purchased, but the shared toilet in the *jhuggi* (slum) was fine. Generating paperwork was expensive. This paperwork included identity cards, residency proofs, income certificates, and other essential documents for accessing government services, employment, education and various entitlements. Generating this paperwork was time-consuming and financially demanding, with several interconnected factors contributing to the overall cost.

First, throughout the interviews, it was apparent that generating, recovering or maintaining documents and submitting them through the proper administrative channels requires hiring a third-party fixer to run the documents to the district administration headquarters or some other administrative office. Such fixers might save time because they are already familiar with the offices and how forms must be completed to be considered and accepted. They have the linguistic and even local connections to grease the process in favour of their client. However, these services entail embedded costs, such as document issuance, notarisation and photocopies, which can strain the limited budgets of the urban poor and migrants, often amounting to thousands of rupees. Such intermediaries often take the form of NGOs; however, the vast majority are not charity outfits but for-profit actors who can exploit the vulnerability of their clients or contribute to what are commonly described as employment-related debt traps (Deshingkar 2022).

Kalpana (age 26), a young mother, described the horror of having her documents stolen and the two months she spent obtaining new ones from offices and schools. She had come to Kapashera seven years earlier to marry, staying at home with her children while her husband worked in a garment factory.

The informality of work arrangements and housing contributed to necessary fudging and creative solutions that demanded a fixer – migrant housing arrangements are typically oral agreements, with no paper trail beyond perhaps a utility payment. Pursuing formal work may depend on paperwork that is therefore unobtainable. Suman, a woman in her forties from Uttar Pradesh, recalled:

> We even had the labour card,[2] made some one and a half years back. That is gone, too. We didn't receive anything. They charged us 8,000 Rs for making it, but we did not get anything.

She continued:

> When they were giving money during COVID, the office worker called me and asked me to send my PAN and Aadhaar card details, and it would get done. But I still haven't gotten it. It cost me INR 400–500 to get it done, but I didn't receive it. (Interview, translated from Hindi, 8 August 2023)

Other costs included the price of transportation, including travel to one's village to retrieve records, if they existed. The time spent obtaining or failing to obtain supporting documents such as birth certificates or proofs of address and income, plus lost wages for days spent in pursuit of records and filing procedures, also racks up opportunity costs for migrant workers, adding to the financial stress that entitles them to government assistance in the first place. This process also diverts resources from other expenses such as food, education and healthcare. Suman noted in relation to her mobile phone:

> It has become expensive to recharge it. A three-month pack costs INR 700. If you have five phones, then it would cost a salary to run it. (Interview, translated from Hindi, 8 August 2023)

More abstract is the cost of civic inclusion and representation. As many explained, their lack of residency made voting impossible. Voting was an option for some who were only a few hours' bus ride away from home. For many, voting at home was also a moot gesture without the potential benefits of village machine politics in their everyday lives. Often, interviewees claimed indifference to politics altogether. Kalpana disavowed knowledge of politics, claiming to rely on her family to make her political decisions. *'What do I know about politics? I push the button that my in-laws tell me.'*

Here, the lack of proper documentation leads to financial burdens that significantly erode the possibility of political representation on the informational periphery. The inability to secure necessary paperwork not only traps individuals in cycles of economic precarity but also marginalises them from the very processes that could help them advocate for their rights and improve their circumstances. This exclusion from urban citizenship, in turn, negatively impacts economic stability. Without proper documentation, migrant workers like Kalpana and Suman are excluded from government benefits, healthcare and educational opportunities, perpetuating their economic vulnerability.

Surviving the informational periphery: the pandemic's effect on the use of digital platforms

We sat in an empty classroom, sharing a lemon soda, bare feet balanced on the base of the desk, as Abhay (early twenties) talked about what he had taught himself during the pandemic lockdowns. He had much to say about all things digital – less to report about accessing the PDS. During lockdowns, Abhay had scrolled through social media reels and realised he could get books to read via his device. He had watched a YouTube video about downloading a pirated book in PDF. He smiled, sharing a relatable story of downloading a dozen free books and managing 'to read a few pages of them, too'.

Abhay had become his family's digital record keeper and an expert on 'digital hygiene', he said. Digital hygiene refers to the practices and habits that individuals and organisations adopt to maintain the security, privacy and overall health of their digital information.

> I keep telling [my parents] what I find out. I tell them ... not to share OTPs and to consult me before doing something. My parents both have phones, but they don't know much about digital security, so it's up to me to tell them.

The COVID-19 pandemic had a profound impact on the lives of India's interstate migrant workers, exposing and exacerbating existing digital divides while prompting significant shifts in how they accessed information, services and resources. On 24 March 2020, a country of over a billion and a half people shut down on less than 24 hours' notice, triggering the mass exodus of millions of migrant workers who suddenly found themselves without income to make rent, pay for or procure food outside the home, and no means to travel, with most roads and public transportation shut down (Goldstein 2022). Abhay's moment of quiet curiosity and time-passing was a rare positive anecdote about the experiences that many faced during the lockdowns.

The pandemic reinforced the centrality of mobile phones in the lives of migrants. Many rely on smartphones for communication, accessing information and staying connected with family and work opportunities via WhatsApp. During the lockdowns, migrants increasingly turned to digital sources for information about the pandemic, government guidelines, healthcare and job opportunities, with some exceptions, such as local radio stations (Goldstein 2022). With physical mobility restricted by containment zones (Naik 2020), and many migrants also the target of religious violence during the swirling rumours of deliberate transmission of the disease by Muslims in Delhi (Beydoun 2020), the virus itself was often perceived as the least of the risks posed to those who broke lockdown orders.

As all our interviewees noted, COVID-19 had increased the use of smartphones and led many households to acquire more than one device to allow multiple children to attend school remotely. Parents, particularly mothers, registered more complaints about using smartphones, devices they had embraced for everyday household needs, including online shopping. A young woman, Chanda, observed that their use could be beneficial 'if you are smart enough to understand. If you aren't, it's a very dangerous thing. I think it's good for the adults, not so much for the kids.'

She had issues with her adolescent son, who previously had not owned a device. After borrowing the money to buy another phone, which she had still not managed to pay back, Chanda's son did not study and used the device for gaming. 'Had he done well, it would not have mattered how much trouble we faced, but he didn't. I ask him to call someone or return the phone to me, but he doesn't' (Interview, 7 August 2023).

The shift to online transactions included increased use of digital payment platforms for sending remittances to extended family, paying bills and making everyday household purchases. Telemedicine and health-related apps gained prominence during the pandemic. Migrants, especially those in remote areas, began using these services to access medical advice and consultations (Kumar et al. 2020).

Indian governments (including individual state governments) introduced and used existing digital platforms to deliver relief measures and subsidies during the pandemic. This required migrants to navigate online systems to access essential resources. However, digital divides persisted. These divides manifest in various ways. In the case of interstate migrants, digital divides have tended to involve limited access to expensive smartphones, stable internet and electricity connections, or the skills, including literacy or multilingual literacy, to use digital platforms effectively (Interview with the Director of Agrasar, 4 August 2023).

While employers often scrambled to keep workers from returning home, the first lockdown triggered massive layoffs in informal sectors and reduced income due to decreased demand in specific sectors (Jesline et al. 2021). As participants in the research reflected, this affected their ability to afford and maintain digital devices and data plans when they seemed most indispensable. This also increased stress and anxiety. Well-being and mental health concerns were often pathologised without consideration for the broader human rights abuses that triggered them (Adhikari et al. 2020). India's linguistic and regional diversity also posed communication barriers to digital access, especially for government services, and paperwork required for entitlements shifted to digital platforms exclusively (Jesline et al. 2021). The COVID-19 pandemic prompted migrants to adopt new digital behaviours out of necessity – access became more critical for information, services and income generation. As use accelerated, however, the pandemic also revealed the dimensions of those disparities in both digital access and skills, sharpening the disparities of an informational periphery. Migrants and their children, like Abhay, who benefited from digital platforms, stood in contrast to those who struggled with digital literacy (often members of the same household), emphasising the uneven landscape of digital inclusion within the informational periphery.

Digital divides within the informational periphery

*How will I use it when I haven't studied? *laughs** – Suman

Suman migrated from Begusarai, Bihar, for marriage. A cheerful woman in her early forties, she still identified as a migrant after 20 years. She was sparse with details on why she had joined her partner on the outskirts of Delhi – it is not uncommon for women to stay back in the village – but she had joked that her new husband was 'having trouble with his meals' and eating out too much. Between his garment factory work and her piecemeal stitching, they sometimes brought in 20,000 Rs a month.

Suman's story highlights the digital divides that exist within the informational periphery related to age and gender. Like most women in the research, Suman did not finish school but was literate in Hindi and knew some English. She used a smartphone, focusing on how the device could advance her children's education. Still young by most standards, Suman and other participants in their forties and fifties spoke of themselves as if from a much older generation. In contrast to Abhay, whose phone never left his hand, a man in his mid-fifties shared that he once dropped his phone in a ditch and watched it float away. He told this story with the stoic grace of someone who had accepted that he was too old to worry about such trivialities.

Throughout the interviews, age and gender gaps appeared to significantly impact facility with devices, digital platforms, and an ability to engage with online forms and paperwork. These disparities reflected broader inequalities: female migrants, especially from conservative regions, may have limited access to education or may not be encouraged to develop digital literacies. However, not all female migrants are wives and mothers, or, if so, are confined to the home. This resonates with what Cowan (2021) identifies as a kind of 'rooted flexibility' that often shapes the daily practices and balancing act of female migrant labour in Gurugram, both in and beyond the home. Many female migrants earn substantially for their families as either heads of households or adult children (Gidwani and Ramamurthy 2018). They often negotiate the informational periphery and its weak urban infrastructures by creatively using technologies to access education and work (Datta (Ayona) 2020).

Nonetheless, participants in this research, such as Suman, often relied on male family members or intermediaries for assistance in managing digital records and online forms. This raises the issue of who manages the household pursuit of PDS rations and other subsidies when it requires skills that are more common among teenage and young adult children. While local NGOs have offered gender-sensitive digital literacy training that consider the needs of women with less formal education than their male counterparts, as well as various domestic responsibilities that require special scheduling, such offerings involve challenging existing gender norms that prevent women from pursuing further education, mobility or autonomy through a standard skills-building approach.

Age and gender gaps in digital literacy reflect broader disparities, but further study is needed to unpack how the intersections of age, gender, education and sending region play out. Furthermore, the variation in how different groups use digital technologies makes it problematic to insist on standardised conceptions of such literacy – it may be more generative to consider many iterations or 'techno-literacies' that have evolved in response to the obstacles of the informational periphery.

NGOs as intermediaries on the informational periphery

NGOs have worked for decades to address age and gender gaps in technological literacy and paperwork among interstate migrants (Goldstein 2022). Initiatives have emphasised skills development as a portable asset for migrants entering the labour market (Interview with the Director of Agrasar, 4 August 2023). CBOs and local entrepreneurs also play a crucial role, creating a shadow PDS of care and assistance. This shadow PDS helps migrants access digital state services and offers a solution for those who fail to access those benefits through official channels.

Located in Sikanderpur village on the edge of Gurugram, Agrasar is difficult to reach by car. The village lanes are narrow enough for an autorickshaw to pass, but cabs usually drop me at the entrance. Over the years, the NGO has grown from a small classroom to occupy several storeys of a new building. Like the NGOs in Kapashera, Agrasar is within walking distance of the *jhuggis* and tenements where children and youth can easily attend a class or volunteer. Migrant-serving NGOs like Agrasar help their beneficiaries in many ways while conducting research (occasionally commissioned) to answer demographic questions about the thousands of migrant households who go uncounted in census figures.

Many migrant-serving NGOs work to help their beneficiaries navigate the informational peripheries of life in an urban slum. However, in recent years, many have seen a shift in their primary activities.

During the pandemic, Agrasar engaged in emergency interventions for thousands of families, helping migrants access food, train tickets and legal aid through WhatsApp. Legal aid was critical for recovering promised or back-wages after the layoffs of the pandemic lockdowns in 2020 (Goldstein 2022). Post-COVID, priorities shifted again. While education had always been at the core of its mission, Agrasar had moved away from its former skills development workshops and begun a curriculum designed to target a specific suite of digital government platforms, namely the 'One Nation One Ration Card' scheme. The related applications are designed for a single user online, but the presence and support of such NGOs suggests a certain weight to these web pages. They are not to be approached on their own but require training and preparation.

These sites can go down, sometimes for a month or more. Agrasar's director and co-founder explains this over chai in his office between staff meetings. The ration kiosks are only open at certain times of the month. In the days before the pandemic, such intermediaries, or 'facilitators', acted as fixers, working on various administrative tasks, helping clients access residency status, and dealing with access to healthcare and the PDS.

Informal networks at the community level also help migrants understand the paperwork involved in PDS enrolment. The classrooms and offices used for conducting research also doubled as tutoring spaces, where free internet access and support could be marshalled from whoever was minding the office desk to help access resources or fill out online forms. Several participants, particularly male youths, mentioned they were doing 'document work' for extra income or planning to start a centre in their home village to facilitate the process of building digital hygiene for migrants in anticipation of a move to the city.

However, Agrasar's director stressed his concerns about the functionality of the sites. When a website crashed, a small broken link icon gave no room for objection. Contact numbers were never answered. 'There is no webmaster,' he said, painting a picture of an empty control centre in which the fate of millions of Indian citizens accessing the PDS had fallen to the desk of an unnamed worker on an indefinite lunch break. Without access to that individual, one might be able to harass a local official. There were few recourses but to wait or carry on without assistance.

Conclusion: migrants and the persistence of bureaucratic exclusion

This chapter has documented the experiences of India's interstate migrants who navigate the transition from 'paper truths' to digital authentication, living on both the informational and geographical peripheries of cities such as Delhi and Gurugram. The ethnographic insights reveal the practices and practical challenges these communities face – challenges that the digitisation of government programmes alone cannot solve. While digital platforms could ostensibly streamline bureaucracy and reduce the high costs associated with obtaining and maintaining official paperwork, they also replicate and sometimes worsen the barriers they seek to eliminate.

Ethnographic observations from Kapashera underscore the paradoxes of digital authentication. While digital tools promise savings in both time and money, the expense of accessing these technologies often negates potential financial benefits for households. This chapter demonstrates that although digital tools and platforms offer new methods of managing identity and citizenship, they also perpetuate the challenges of paper-based systems. This includes the ongoing struggle for migrants to authenticate their identities and access basic services amidst a web of bureaucratic red tape that has merely shifted from the tangible to the digital. The persistent reliance on both paper and digital records raises questions about the rights of citizens as record bearers and digital subjects – whether these records are stored on paper or in digital databases. For example, the debate around the 'Right to be Forgotten' in Indian law and digital privacy further highlights the complexities of this transition. Although India does not have specific legislation for the right to be forgotten, the concept is becoming increasingly relevant in legal discussions, especially with the rising importance of digital data and privacy (Alok 2017; Siddiqui 2022).

Furthermore, informational peripheries are not just geographic margins but also conceptual spaces where migrants encounter barriers to accessing and utilising technology. The digital turn has not dissolved these peripheries but often reinforces them through new forms of exclusion and bureaucratic opacity. In other words, the shift from paper to digital documentation does not eliminate the structural violence embedded in these processes but transforms its modalities, embedding new forms of marginalisation in the digital landscape.

The notion of paper truths, as articulated by Tarlo (2001; 2003), provides a lens to examine how documentation, whether on paper or via a digital interface, serves as a tool of power and control. The historical continuity of red tape, from colonial and postcolonial-era paper to the current shift to digital, underscores the enduring nature of this alternative truth – the one recognised by the state bureaucracy on the page or screen – and its everyday power over those who live on the informational periphery. This power dynamic illustrates how bureaucracies perpetuate a form of control that exacerbates the peripheralisation of vulnerable populations within physical and informational spaces.

NGOs like Agrasar have become lifelines in bridging the paper truth/digital divide, emphasising the ongoing need for human intermediaries. However, their efforts also highlight the gap between the theoretical promises of digital platforms and their practical realities. While an absentee webmaster may have replaced the local bureaucrat, the role of the middleman has remained. This continuity of intermediaries signals a persistent need for human intervention, reflecting the inadequacy of digital platforms to address the deeper systemic issues within bureaucracies.

Ultimately, the informational periphery exists not only in the spatial geography of a city but within peripheralised groups, such as interstate migrants. The possibility of differentiation within a periphery raises critical questions for further consideration: can we understand the informational periphery without grappling with what indeed constitutes the informational *centre* of a city or community, and to what extent are individuals in this so-called centre also peripheral in their understanding of broader realities in the city? As we look to the informational periphery as a framework for grappling with knowledge gaps among the poor, we find the periphery an ever-moving target – there are peripheries within peripheries. In this chapter, both gender and age differences in literacies and access highlight the need to continually and critically engage with the spatial and social hierarchies that produce, reinforce and internally differentiate informational peripheries within marginalised populations.

Notes

1 The ration card system in India has colonial origins, established parallel to Britain's World War II rationing schemes, though they often floundered at the village or rural scale (Sriraman 2018). Unsurprisingly, the urban–rural disparities of the colonial ration card have not been resolved in their contemporary counterparts, especially among migrants shuttling back and forth between urban and rural areas.

2 The Labour Card, also known as the Shramik Card or eShram, is a government-issued identification document for informal workers. See https://eshram.gov.in (accessed February 2025).

References

Abbas, R. 2016. Internal migration and citizenship in India. *Journal of Ethnic and Migration Studies* 42(1): 150–68. https://doi.org/10.1080/1369183X.2015.1100067.

Adhikari, A., Goregaonkar, N., Narayanan, R., Panicker, N. and Ramamoorthy, N. 2020. Manufactured maladies: Lives and livelihoods of migrant workers during COVID-19 lockdown in India. *Indian Journal of Labour Economics* 63: 969–97. https://doi.org/10.1007/s41027-020-00282-x.

Agarwal, S. and Agnihotri, A. 2022. Food security for interstate migrants: An empirical analysis of the ONORC. *Economic & Political Weekly* 57(51). Accessed February 2025. https://www.epw.in/journal/2022/51/commentary/food-security-interstate-migrants.html.

Alok, P. K. 2017. 'Right to be forgotten' in Indian law. *Economic & Political Weekly* 52(11). Accessed February 2025. https://www.epw.in/journal/2017/11/law-and-society/%E2%80%98right-be-forgotten%E2%80%99-indian-law.html.

Anderson, B. 2006. *Imagined Communities: Reflections on the origin and spread of nationalism*. Rev. edn (orig. 1983). London: Verso.

Bayly, C. A. 2000. *Empire and Information: Intelligence gathering and social communication in India, 1780–1870*. Cambridge: Cambridge University Press.

Beydoun, K. A. 2020. Between virus and violence: The horror of being Muslim in India. *The New Arab*, 17 April. Accessed February 2024. http://newarab.com/opinion/between-virus-and-violence-being-muslim-india.

Breman, J. 1996. *Footloose Labour: Working in India's informal economy*. Cambridge: Cambridge University Press.

Chatterjee, P. 1993. *The Nation and its Fragments: Colonial and postcolonial histories*. Princeton, NJ: Princeton University Press.

Cohn, B. S. 1987. The census, social structure and objectification in South Asia. In *An Anthropologist among the Historians and Other Essays*, 224–54. Oxford: Oxford University Press.

Cowan, T. 2021. Rooted flexibility: Social reproduction, violence and gendered work in the Indian city. *Gender, Place & Culture* 28(1): 66–87. https://doi.org/10.1080/0966369X.2019.1708276.

Datta, Amrita. 2020. Circular migration and precarity: Perspectives from rural Bihar. *Indian Journal of Labour Economics* 63: 1143–63. https://doi.org/10.1007/s41027-020-00290-x.

Datta, Ayona. 2020. The 'smart safe city': Gendered time, speed, and violence in the margins of India's urban age. *Annals of the American Association of Geographers* 110(5): 1318–34. https://doi.org/10.1080/24694452.2019.1687279.

Datta, A. 2024. The informational periphery: Territory, logistics and people in the margins of a digital age. *Asian Geographer* 41(2): 125–42. https://doi.org/10.1080/10225706.2023.2253233.

De Wit, J. and Berner, E. 2009. Progressive patronage? Municipalities, NGOs, CBOs and the limits to slum dwellers' empowerment. *Development and Change* 40(5): 927–47. https://doi.org/10.1111/j.1467-7660.2009.01589.x.

Deshingkar, P. 2022. Cultural capital and constrained agency in debt-migration for construction work in India. *Cultural Studies* 37(4): 605–25. https://doi.org/10.1080/09502386.2022.2045333.

Dirks, N. 2011. *Castes of Mind: Colonialism and the making of modern India*. Princeton, NJ: Princeton University Press.

Dixon, P. 2017. A failure to 'do no harm': India's Aadhaar biometric ID program and its inability to protect privacy in relation to measures in Europe and the U.S. *Health Technology* 7: 539–67. https://doi.org/10.1007/s12553-017-0202-6.

Economic & Political Weekly Engage. 2020. Food security and COVID-19: Why India's public distribution system requires an overhaul. 19 June. Accessed February 2025. https://www.epw.in/engage/article/food-security-and-covid-19-why-indias-public.

Gidwani, V. and Ramamurthy, P. 2018. Agrarian questions of labor in urban India: Middle migrants, translocal householding and the intersectional politics of social reproduction. *Journal of Peasant Studies* 45(5–6): 994–1017. https://doi.org/10.1080/03066150.2018.1503172.

Goldstein, S. 2022. Urban exodus, mobility injustice. *Places Journal*, March. Accessed February 2025. https://placesjournal.org/article/urban-exodus-and-mobility-injustice-in-india.

Gooptu, N. 2019. JSAD special issue on skill development in India. *Journal of South Asian Development* 13(3): 241–8. https://doi.org/10.1177/0973174118822391.

Gupta, A. 2012. *Red Tape: Bureaucracy, structural violence, and poverty in India*. Durham, NC: Duke University Press.

Henne, K. 2019. Surveillance in the name of governance: Aadhaar as a fix for leaking systems in India. In *Information, Technology and Control in a Changing World: Understanding power structures in the 21st century*, edited by B. Haggart, K. Henne and N. Tusikov, 223–46. Cham: Springer/Palgrave Macmillan.

Hindustan Times. 2020. Haryana launches online portal to facilitate return of migrant workers. 3 May. Accessed February 2025. https://www.hindustantimes.com/gurugram/haryana-launches-online-portal-to-facilitate-return-of-migrant-workers/story-cJ5Bklh1c8qZhxrNuVP8HO.html.

Hirudayaraj, M., Barhate, B. and McLean, G. N. 2024. Work conditions of interstate migrant workers in India: A critical realist exploration. *Human Resource Development Quarterly* 35(4): 431–53. https://doi.org/10.1002/hrdq.21515.

Hull, M. 2012. *Government of Paper: The materiality of bureaucracy in urban Pakistan*. Berkeley, CA: University of California Press.

Jayaram, N. and Varma, D. 2020. Examining the 'labour' in labour migration: Migrant workers' informal work arrangements and access to labour rights in urban sectors. *Indian Journal of Labour Economics* 63(4): 999–1019. https://doi.org/10.1007/s41027-020-00288-5.

Jesline, J., Romate, J., Rajkumar, E. and George, A. J. 2021. The plight of migrants during COVID-19 and the impact of circular migration in India: A systematic review. *Humanities and Social Sciences Communications* 8: 231. https://doi.org/10.1057/s41599-021-00915-6.

Kantor, H. 2018. Building beyond the bypass road: Urban migration, ritual eating, and the fate of the joint family in Patna, India. *American Anthropologist* 120(2): 212–23. https://doi.org/10.1111/aman.12972.

Kumar, K., Mehra, A., Sahoo, S., Nehra, R. and Grover, S. 2020. The psychological impact of COVID-19 pandemic and lockdown on the migrant workers: A cross-sectional survey. *Asian Journal of Psychiatry* 53. https://doi.org/10.1016/j.ajp.2020.102252.

Maddox, J. and Kanthawala, S. 2023. The revolution will be forwarded: Interrogating India's WhatsApp imaginary. *Journal of Communication Inquiry* 47(3): 249–67. https://doi.org/10.1177/01968599221095177.

Medappa, K. 2023. Rethinking mutual aid through the lens of social reproduction: How platform drivers ride out work and life in Bengaluru, India. *Journal of South Asian Development* 18(3): 383–408. https://doi.org/10.1177/09731741231162450.

Mukhopadhyay, A., Bagchi, K. K. and Udo, G. J. 2023. Exploring the main factors affecting mobile phone growth rates in Indian states. *Journal of the Knowledge Economy* 15: 5746–68. https://doi.org/10.1007/s13132-023-01206-y.

Naik, M. 2019. Negotiation, mediation, and subjectivities: How migrant renters experience informal rentals in Gurgaon's urban villages. *Radical Housing Journal* 1(2): 45–62. https://doi.org/10.54825/BXCJ5614.

Naik, M. 2020. State-society interactions and bordering practices in Gurugram's pandemic response. *Urbanisation* 5(2): 181–90. https://doi.org/10.1177/2455747120974531.

Naik, M., Lahiri, A., Kaushik, U. and Manish. 2022. A case for functional social protection portability to address vulnerabilities of migration-affected children. *Economic & Political Weekly* 57(44–5). Accessed February 2025. https://www.epw.in/engage/article/case-functional-social-protection-portability.

Poell, T., Nieborg, D. and van Dijck, J. 2019. Platformisation. *Internet Policy Review* 8(4). https://doi.org/10.14763/2019.4.1425.

Raman, B. 2012. *Document Raj: Writing and scribes in early colonial South India*. Chicago, IL: University of Chicago Press.

Ray, A. 2024. Coping with crisis and precarity in the gig economy: 'Digitally organised informality', migration and socio-spatial networks among platform drivers in India. *Environment and Planning A: Economy and space* 56(4): 1227–44. https://doi.org/10.1177/0308518X231220296.

Routray, S. 2022. *The Right to be Counted: The urban poor and the politics of resettlement in Delhi*. Redwood City, CA: Stanford University Press.

Santalahti, M., Sumit, K. and Perkiö, M. 2020. Barriers to accessing health care services: A qualitative study of migrant construction workers in a southwestern Indian city. *BMC Health Services Research* 20: 619. https://doi.org/10.21203/rs.3.rs-16097/v4.

Shaw, J. and Graham, M. 2017. An informational right to the city? Code, content, control, and the urbanization of information. *Antipode* 49(4): 907–27. https://doi.org/10.1111/anti.12312.

Siddiqui, N. 2022. Claiming the right to be forgotten. *Economic & Political Weekly* 57(3). Accessed February 2025. https://www.epw.in/journal/2022/3/letters/claiming-right-be-forgotten.html.

Singh, R. 2019. Give me a database and I will raise the nation-state. *South Asia: Journal of South Asian Studies* 42(3): 501–18. https://doi.org/10.1080/00856401.2019.1602810.

Singh, R. and Jackson, S. 2021. Seeing like an infrastructure: Low-resolution citizens and the Aadhaar Identification Project. *Proceedings of the ACM on Human-Computer Interaction* 5 (CSCW2): 1–26. https://doi.org/10.1145/3476056.

Sriraman, T. 2018. *In Pursuit of Proof: A history of identification documents in India*. Oxford: Oxford University Press.

Srivastava, S. 2012. Duplicity, intimacy, community: An ethnography of ID cards, permits and other fake documents in Delhi. *Thesis Eleven* 113(1): 78–93. https://doi.org/10.1177/0725513612456686.

Stoler, A. L. 2008. *Along the Archival Grain: Epistemic anxieties and colonial common sense*. Princeton, NJ: Princeton University Press.

Tarlo, E. 2001. Paper truths: The emergency and slum clearance through forgotten files. In *The Everyday State and Society in Modern India*, edited by C. J. Fuller and V. Benei, 68–89. London: Hurst & Co.

Tarlo, E. 2003. *Unsettling Memories: Narratives of the emergency in Delhi.* Berkeley, CA: University of California Press.

Upadhya, C. 2016. Engineering equality? Education and im/mobility in coastal Andhra Pradesh, India. *Contemporary South Asia* 24(3): 242–56. https://doi.org/10.1080/09584935.2016.1203863.

11

Separate informational worlds: urban governance and speculative futures in the periphery of Mexico City

Julie-Anne Boudreau

In February 2020, as the world closely monitored the spread of the COVID-19 pandemic, Mexico City's mayor, Dr Claudia Sheinbaum, asked the city's Digital Agency for Public Innovation (ADIP) – a body usually responsible for data management and bureaucratic digitalisation – to develop an epidemiological model that would be a primary tool in planning hospital needs. In our interview with a high-ranking official, he recalled his conversation with the mayor (Interview 9, ADIP, 2 June 2022). Sheinbaum asked him, the 'tech guy', to develop an epidemiological model – despite the absence of any public health expertise – as the Health Department lacked analytical capacity. For the mayor it was important to develop a practical, digital planning tool to inform decision making. Entrusting this task to the ADIP had the advantage of creating a simple, easy to use model, unencumbered by the complexities of epidemiological models typically developed by public health experts.

The ADIP's first decision, which was an unusual one for epidemiological modelling, was to choose as the main variable hospitalisation rates rather than infection rates. This approach made sense, because the objective of the model was not really to analyse epidemiological behaviour, but to plan for hospital capacity. Every night at 9 p.m., a team of 30 employees called the director of every single hospital in the city to collect data on four numbers: new hospitalisations, deaths, general hospitalisations, and hospitalisations requiring ventilators. With these four simple variables, the ADIP created a couple of scenarios based on what the official called 'public policy shocks', by which he meant the anticipated effects of government interventions such as lockdowns or mask mandates on infection rates. He explained:

> I am an economist, these are relatively simple partial differential equations and what matters is the determination of the parameters, everything else becomes mechanical after that. But the important part here for us is that most of the models defined their parameters based on the literature, which is the only thing we had: from Wuhan they told us what was the number of days spent intubated and with that how many resident days, etc. But for me the important thing was to think about what we were going to alter; the only thing we could alter as a government was the reproductive rate in T, in T1, or whatever. So, what we did was to say to the mayor: OK, today it looks like this. If we don't do anything, the curve looks like this, but it's not that we're not going to do anything. One day we're going to close down. What do we expect to happen if we close? (Interview 9, ADIP, 2 June 2022)

This curve became Mexico City's most important decision-making tool between 2020 and 2022 because it produced scenarios predicting where things would go if the government introduced 'public policy shocks'. The official explained: 'For me, the curves were useful to know if what we were seeing was in line with what we thought we were going to be able to do to influence in reality' (Interview 9, ADIP, 2 June 2022). This interview highlights the importance of decision-making tools in emergency situations, as they provide a framework for guiding governmental responses.

In that sense, the objective of this chapter is to examine state responses to the COVID-19 pandemic by analysing these 'public policy shocks' in the context of the 'informational world' in which the state was acting – a world that had been shaped over decades by various informational practices for planning and decision making. My focus will be on decision-making tools, particularly computer modelling of the spread of the COVID-19 virus, to draw a picture of how state actors perceive peripheral urban realities. This analysis draws on seven extensive interviews with civil servants from various policy sectors (health, transport, social services, data management) who were involved in managing the pandemic in Mexico City, as well as a review of policy documents.

This central city-state informational world will be contrasted with the informational world of both borough-level civil servants, who deployed what was called 'territorial knowledge' in our interviews, and women living in peripheral areas of the city. In May 2021, we conducted 14 hours of creative mapping workshops via Zoom with two groups of women. The first group had the opportunity to work from home during the lockdown, while the second group continued working outside the home in various jobs such as domestic work, beauty services, or selling cakes or cosmetics. We saw each group three times, for approximately three hours per session. During these sessions, we discussed how they related to the state's COVID responses, their fears and hopes, their routines, and their response strategies. We mapped their domestic spaces, their routes if they worked outside the home, the places they would go if they could, and their bodies. Drawing on these creative mapping workshops with 21 women, the objective of this chapter is to analyse their responses to the pandemic. This will reveal a huge gap between the state's understanding of peripheral urban realities through digital mapping tools and modelling, and the lived realities of women, as expressed through their participatory creative maps.

By 'informational world', I mean an ensemble of practices of data gathering, interpreting and analysing based on a shared imaginary of what constitutes legitimate and reliable information, knowledge and 'truth'. While Kitchin and colleagues (2015) speak of 'data assemblages', I argue that the concept of 'world' better captures the ontological and epistemological assumptions that shape decision making.[1] Following Nancy (1993), this chapter suggests that each informational world is characterised by a specific sensory experience and by a shared space of understanding. The informational world of city-level state actors is very different from, and more hegemonic than, the informational worlds of both borough-level employees who work with 'territorial knowledge' and peripheral residents. In short, these residents and civil servants operate in what Datta (2024, 4) calls 'informational peripheries' – that is, informational worlds that are 'both geographically and digitally distant' from the central city-state informational world.

This comparison of informational worlds during the pandemic in Mexico City will show that both borough-level civil servants and peripheral inhabitants create urban futures more through 'repair' than through digitalised models (the core of the central city-state informational world). The chapter begins by theoretically framing repair as the main characteristic of peripheral informational worlds, and how this leads to speculative futures. I then turn to analysing the decision-making process during the pandemic in order to draw a picture of the hegemonic informational world of the state, which becomes particularly evident in moments of crisis management. The chapter then dives into the informational worlds of the peripheries of Mexico City during the pandemic, looking first at borough-level civil servants and then at residents. In conclusion, I will return to the notion of repair and speculative futures in the informational periphery.

Repair and speculative futures in peripheral informational worlds

The literature on repair has three distinct origins. First, science and technology studies and infrastructure studies suggest analysing technological development through the lenses of failure, repair and maintenance. Graham and Thrift (2007) propose a focus on moments of infrastructural breakdown as the key to understanding innovation – what Jackson (2014) calls 'broken world thinking'. In this body of work, repair is defined as 'subtle acts of care by which order and meaning in complex sociotechnical systems are maintained and transformed, human value is preserved and extended, and the complicated work of fitting to the varied circumstances of organisations, systems, and lives is accomplished' (Jackson 2014, 222). This is a fundamentally different theoretical basis from that of planning and modelling, requiring a shift from a language of use and function to a language of distributed agency, decentred from human control (Jackson and Kang 2014).

Repair is not to be imagined as something pertaining exclusively to non-state informational worlds. On the contrary, it is a mode of doing and being that is dominant within the state as well, but it is important to distinguish *where* in the state. For example, Alejandro De Coss-Corzo's (2021) study of water repair workers in Mexico City shows how, in the absence of accurate or updated plans of the grid, workers develop embodied, creative and practical knowledge to repair leaks. He notes: 'Their goal there is not to return things to their ideal or original form, but to fashion normality as an ongoing process that is always shaped by infrastructure and other socio-material relations' (De Coss-Corzo 2021, 247). This 'logic of adaptation' characterises the production of urban space in Mexico City more generally. As De Coss-Corzo concludes, 'against ideals of standardised infrastructure, it is patchwork what enables infrastructure to function. It does so not through thorough planning, but through a calculation of probabilities and the deployment of ways of knowing and doing based on previous practice and experiences' (De Coss-Corzo 2021, 247). What this example demonstrates is that the 'state' functions through a variety of informational worlds: that of the engineers who plan the water system or the command centre managing the pandemic is quite distinct from that of the water repair workers or the borough-level 'territorial team'.

A very different body of work on repair comes from the anthropology of suffering. Exploring massive episodes of social suffering caused by wars, disasters or colonialism, anthropologists describe how people cope, recover, heal and rehabilitate (Das et al. 2001; Harrison 2019). This approach comes closer to the notion of care, especially if care is understood as relational and interdependent. For instance, Puig de la Bellacasa (2017) calls for a speculative understanding of care, by which she means being attentive to possible futures that can emerge from a situation. Speculative thinking is a mode of living that works beyond human intentionality. It requires care as a living involvement in the situation, not as an idealised normative obligation. In short, caring is a living technology, a co-transformation in an interdependent web. What I wish to argue with the example of the informational world of pandemic management in the periphery of Mexico City is that this speculative (and improvisatory) form of care and repair is perhaps more productive than the disembodied planning logic of the central-city informational world.

A third body of literature on repair, which is very important in Latin America but perhaps less directly relevant to this chapter, is grounded in legal and human rights studies, looking at transitional, restorative and reconciliatory justice. This is a vivid debate in the context of transitions from state violence and dictatorships to democratic regimes in Latin America. Judicially, these debates affirm the right of victims to know the truth, to repair loss and damage, and to receive a guarantee that the traumatic event will not be repeated. Confining repair to the judicial arena can, however, be criticised. Studying the judicial process of repair and compensation for the Bhopal disaster in 1984 in India,[2] Veena Das shows how victims were forced to 'transform their suffering into the language of science in order for it to be judicially recognised' (Das 1997, 569). Again, in this body of literature, repair is entangled within differentiated informational worlds. What is considered 'hearable' or 'useful' information in the judicial arena is very different from the information about suffering that victims may want to provide.

In short, informed by these three definitions of repair, I mean the repair of infrastructural failures, but also (and perhaps primarily) how everyday 'minor gestures' (Manning 2016) are constantly enacted as a consequence of massive infrastructural failures. Repair is a survival strategy that generates speculative rather than planned futures (Guma et al. 2023). Bryant and Knight argue that speculation is what flourishes in the gap that emerges from 'shattered expectations, an inability to anticipate, and a lack of historical anchors' (Bryant and Knight 2019, 79). In other words, when the world seems to slip from the confines of our informational world, when information becomes too partial, conjectural and confidential, then people 'attempt to construct a meaningful interpretation by pooling their conversational resources' (Bryant and Knight 2019, 94). This is how peripheral informational worlds work.

The informational world of a central-city pandemic command centre

There is ample literature on the management of the pandemic in cities, particularly focusing on its reliance on technologies and mapping. While many studies focus on the role of surveillance and monitoring technologies (Datta et al. 2021; Sonn and Lee 2020), in Latin American countries the state does not have the capacity to implement extensive digital surveillance systems that can be found in other parts of the world (Waisbord and Segura 2020). In the case of Mexico, the state managed the crisis without smart city technologies, or what Ananth (2020) labelled COVtech. Instead, as I will describe in a moment, decision making relied on human labour to power monitoring and modelling instruments. Moreover, the focus was not on surveillance, but on planning. During the lockdown, police officers in Mexico City had no power to fine people or control their movements; their task was to convince them to comply. This entailed an informational world characterised by decision-making tools very similar to those used outside of emergency management: modelling, mapping, planning. These tools work on a linear temporality, are not responsive in real time, and involve a much smaller volume of data.

For instance, infection intensity and high-contagion neighbourhoods were identified through the extensive use of maps. In order to analyse the spread of the virus and its geographic distribution, the government reproduced at the national scale the dashboard created by Johns Hopkins University, where cumulative cases, deaths and the progression of vaccination were displayed and updated every 24 hours.[3] The aim of this spatiotemporal map was to support the government in identifying the most affected areas and developing prospective models. Beyond these aims, as a number of critics have argued, these dashboards produce strong affective and behavioural effects (Kent 2020).[4] Andrea Pase suggests that this online cartography 'allows you to see the virus spread, yet it remains neutral, a buffer; it spectacularises the dramatic events, and it works like a screen onto which emotion, fear and expectations can be projected' (Pase et al. 2021, 137).

In what follows, I will closely analyse these modelling practices as the core of the central-city state informational world. Maps, graphics, tables and numbers are the basic units of this informational world. Such 'hard' data, data that 'doesn't lie', data that can be abstracted and generalised, provide decision makers with the confidence to act. In this informational world, decision makers seek ways to simplify an urban reality they know is extremely complex and difficult to read. Beyond the basic principles of simplification and miniaturisation that have been so forcefully analysed in Scott's *Seeing Like a State* (1998), these practices not only provide legibility, they also mostly produce legitimacy.

A clear example of this can be seen in the use of the curve generated by the epidemiological model created by the ADIP. The curve was constantly referenced by the president and all public figures in their information briefings. It served as much for decision making as it served as a pedagogical and communicational instrument. This was the case when, on 27 March 2020, in his daily briefing the president displayed an 'old-fashioned' print version of the curve, using it to reinforce his argument for a lockdown on the basis of the need to introduce 'public policy shocks'; if nothing was done, the situation would be catastrophic.

No matter how 'truthful' it was believed to be, the curve induced immense stress among policy makers. If a policy shock was not introduced in time, when the curve was just beginning to rise, it was already too late. In other words, this curve became a tool for 'seeing' the future. But once the point of no return was reached, there was no way to rectify things and Mexico City would be left without enough hospital beds. During our interview, with tears in his eyes, the civil servant who created the model recalled the peak of hospitalisations on 19 January 2021:

> I was biting my nails, but, as I always told those who got nervous, it's going up and you say: I'm still within the model. What stresses you is that it doesn't stop rising when you project the rise, because when you realise that it's not going to go down, it's often too late. So, the big fight at that moment, not the big fight, the big thing that I was doing all day long, with everybody, was: 'Trust the process, Zen, Zen, do you trust this?' (Interview 9, ADIP, 2 June 2022)

'Trust the process, Zen, Zen' – these abstract projections were the only thing the government had to assess whether they had enough beds with ventilators. The epidemiological model had one simple objective: plan ahead for what was projected to be hospital demand at the peak of infections.

When the Sheinbaum administration took office in 2018, they knew they were arriving after decades of neoliberal cutdowns in public services, and consequently a dilapidated public health infrastructure. Their goal, as the Secretary of Health told us, was to 'stabilise' the public health sector:

> A process of dismantling of services, fragmentation, very little investment, a lot of obsolete equipment, and above all, and something that in general is very difficult to transform quickly: an absolutely corporate, clientelist and business governance. So, we arrived at the Department and what we had was each space working a little bit in the dynamics of those particular interests, and attending to them, but without that being the centre of the Secretariat's work. So, we arrived to stabilise this. (Interview 12, SEDESA, 7 June 2022)

The immediate response when the pandemic hit was of course to increase and strengthen hospital capacity. But in the peripheries, what was even more immediate and important was the complete loss of livelihood for the more than 60 per cent of inhabitants working in the informal sector, mostly street vending. Additionally, the lack of fibreoptic infrastructure in these areas, combined with the fact that most households do not have access to computers, made online schooling impossible. Although the Welfare and the Economic Departments attempted to respond to these socio-economic emergencies, the aid on offer did not reach the great majority of inhabitants, nor was it effective or adequate to address real needs. This can be attributed to two simple reasons: (1) the intended beneficiaries of these programmes were unaware of their existence or did not have the means to claim them (such as internet access); and (2) the civil servants who designed these programmes from the central city's offices on the basis of census data were unaware of the lived peripheral realities. They designed their aid programmes on the basis of past realities abstracted in a census, not on the basis of experiential knowledge of the pandemic's peripheral reality.

Despite this imaginary truthful, objective information based on abstract knowledge, in my conversations with civil servants I realised that in this informational world, a reasoned discourse of objectivity does not preclude intense embodiment of the information. As the director of ADIP exclaimed as he finished explaining how he had developed the epidemiological model:

> It's the clearest geometric shape in my head for my whole life, I have the shape of the hospitalised in the city tattooed under my skin, because I have seen it every day for two and a half years. (Interview 9, ADIP, 2 June 2022)

This is a striking statement, which needs to be understood in the context of many tears shed during the interview, where it became clear that the civil servant was giving himself entirely to his role. In this phrase, he literally embodies the 'abstract and objective' information provided by the curve. This total immersion in their work was also expressed by other interviewees in various ways. While this informational world rests on an imaginary of distance and disembodiment, the experience of producing this knowledge was definitely embodied. The difference between the informational world of borough-level employees and that of peripheral inhabitants is one of language. What is embodied in the city-centre command group of civil servants are curves, numbers, and bird's-eye maps. As we will now explore, in the peripheral informational world, what is embodied are street-level experiences.

The informational world of peripheral pandemic management

In January 2021, as hospitalisations reached their peak, there were 11 beds left in the city. The government responded by sending the least critical cases home. This is when the Social Welfare Department took the lead. Rather than relying on the ADIP's abstract curve, the city now depended on a Citizen Participation Team. Used to walking the city to deliver social services, this 'territorial team' organised home care services and the delivery of oxygen. Once fast-testing services were approved, these too were deployed by the team: 'And we had these groups, which were basically like a hundred mobile units, which were this Participation Team of people who took care of people in the territory, plus the public health team, they did the tests. It was quite the spectacle' (Interview 9, ADIP, 2 June 2022).

Dressed in green vests or white suits, these borough-level employees knew the streets. Their work was grounded in territorial knowledge of the city, not numbers and curves. An area director responsible for street market regulation at the borough level remembers how these territorial teams were deployed locally:

> We patrolled on foot, on horses, in normal patrols. And what we did was that together with our colleagues from the territorial area, teams were put together ... I think it was well designed, because we all had our (hand-written) logs of where we had to go, what we had to do, how we had to do it, what times we had to do it, and who was going to be present that day. Obviously, we were given equipment: mask, gloves, and we wore a white suit, and from there we were called ghostbusters. (Interview 14, Tianguis Iztapalapa, 10 June 2022)

The contrast in methods is important: hand-written logs, street patrols, and workers in white suits, gloves and masks. However, these territorial teams were responsible only for logistical organisation and implementation, not decision making. They responded to decisions coming from above, made on the basis of curves and digital modelling. For instance, the first thing these local 'ghostbusters' or 'green vests' were asked to do was sanitise buses and subway cars when they returned to their peripheral terminals. Their planning and decision-making capacity based on territorial knowledge of the street was not recognised. Instead, they were used as bodies to get the work done and operationalise the decisions made in the mayor's cabinet.

In order to make rapid, informed decisions on managing the pandemic, high-level civil servants depended on abstract, totalising knowledge: map, flows, curves, numbers. This was their lens to calculate what was coming next – this was their entry point into the future. However, what these decision-making technologies lacked was the nuances that remain invisible to numbers and bird's-eye perspectives. The fact that most residents in the eastern periphery worked in the informal sector meant that they continued to be mobile during the lockdown. The model's assumption that the propagation rate would drop from 2.8 to 0.99 if the 'public policy shock' of lockdown was introduced failed to account for geographic variation. As one civil servant from the borough of Iztapalapa, on the periphery of the city, told me:

> In Iztapalapa we have 21 public markets. We have 134 concentrations [of informal street vending], and 354 [formal] street markets ... Iztapalapa is a world of commerce: wherever you stand there are

> traders, that is really our normal work, what is done every day, between the supervision, the procedures that are carried out, that is how we work. (Interview 14, Tianguis Iztapalapa, 10 June 2022)

How can a propagation rate drop from 2.8 to 0.99 on those streets? This is where her description of the teams' patrol work revealed constant creative solutions, adaptation, and what I would like to call repair, understood as a caring response to suffering.

> I mean, it was very complicated and difficult, because we could not make the merchant understand, 'don't come' [during the lockdown]. Well, it's his sustenance, how can I leave you without working for so long. In the merchants' associations, they themselves created ways for people to be more or less in a good economic condition to be able to go on. Some of them sold from inside their homes, so that there would be no shortage. I think that something we should applaud is that there was no shortage [of food]. (Interview 14, Tianguis Iztapalapa, 10 June 2022)

These gestures of care and repair do not appear on the curve or on the map. This begs the question: can other forms of mapping and territorial data collection also serve as decision-making and planning technologies? If so, how? This would imply transforming what we consider solid 'evidence': from abstract to material knowledge, from universal to idiosyncratic perspectives, from fixed to dynamic and interactive tools, from bird's-eye to street-level perspectives. If such mapping techniques generated from a peripheral informational world were brought to central-city command centres, what decisions could be made to respond to a crisis?

The city's territorial teams possess a specific type of knowledge. They know an infinite amount of detail about their territory. They know specific people, street corners, smells; they know what happened to Mrs Maria last week, or where to buy the best quesadillas. They know who's best at repairing watches, where the friendliest shoe shiner is posted, where the street is full of potholes, and where a local politician gained votes because all the streetlights are functioning. Within the city governmental structure, these teams are usually deployed to foster civic participation. They set up information kiosks or provide home services; they also knock on doors to transmit government information. They are recognisable because of their uniform (in the central city administration the uniform consists of green vests emblazoned with the city logo).

Territorial knowledge emerges from inhabiting certain places – from practising in these places regularly. As Escobar (2018) suggests, territories are 'pluriverses' that everybody participates in making. They are 'endowed with capacities, missions, and responsibilities':

> As a collection of people residing and working together with their environments to provide the basics of a life worth living, territory configures porous boundaries that – far from being markers of definition or defence – establish the means for accessing a larger world while curating an experience of coherence for those who live within them. (Simone and Castán Broto 2022, 775)

As a 'means for accessing a larger world', territories generate knowledge that is very different from abstract curves and digital data. Territorial knowledge is of its very essence experiential knowledge; it requires embodiment and lived experience.

Those who produce territorial knowledge, among them the city's territorial teams, can work with an impressive amount of detail. From the outside, particularly from the perspective of civil servants working at computers, these details can be read as 'noise'. They are rarely legible for those who do not participate in the territorial pluriverse. The challenge, therefore, is how to decide what to pay attention to. As Dalton et al. (2016) suggest, data is always spatial, not only in its content, but also because it is always originated and used somewhere. Territorial knowledge originating from the periphery is generally considered unimportant, noisy and messy, if it even reaches the central city. Sitting behind their desks, civil servants do not have the skills to decode such knowledge, because the form that territorial knowledge takes is narrative, idiosyncratic, overloaded with detail, colours, smells. It is not as slick as a curve.

Following Datta (2023), this territorial knowledge presented narratively and messily could be defined as an 'informational periphery' – a bypassing of forms of knowledge that the powerful do not know how to read. This is why it is important to search for ways of mapping territorial knowledge so that decision makers can read this type of data. Listening to territorial teams is one way to begin, because these teams work from what I have called elsewhere a 'street epistemology' (Boudreau 2022). A street epistemology implies describing the articulation between bodies, streets and buildings. It begins with the analysis of street scenes, articulated through mobile bodies moving between situations; bodies that are constantly transiting inside buildings and out on the street.

If we follow this approach, a street epistemology will lead us beyond borough-level territorial teams. This is why we also wanted to explore more closely the experience of peripheral residents and their responses to the crisis. Given their key role in articulating the domestic space with the street and with governmental representatives, we decided to focus on women (Ticktin 2021; Watson et al. 2021).

Our goal with the creative mapping workshops was to understand the women's experiences of the pandemic and explore different scales and methods of mapping. In particular, we were convinced that pandemic decision making should be based on domestic maps: on what happened in people's homes, within their walls – in other words, those spaces that are never mapped. There are statistics documenting a rise in emergency calls related to domestic violence during the lockdown.[5] There are also statistical reports on the effects of the pandemic on school-age children, economic losses, and so on. However, these statistical reports were produced post facto; they serve to analyse the effects of the pandemic, not to make decisions in the midst of a crisis. They do not offer projections and scenarios for action like the epidemiological model did.

The mapping workshops were not originally intended to produce information for decision makers. The maps we created in 2021 were intended primarily to open a healing space for the women who participated (they expressed great gratitude for the opportunity to share their experiences during the workshops), and also for academic audiences. However, as we were analysing the rest of our data, in particular our interviews with civil servants and the programmes they implemented, it became clear that what we first saw as state negligence in the peripheries would probably be better understood as state misunderstanding of the periphery. And this is due to the lack of embodied understanding of peripheral realities. It is not that state representatives ignored the peripheries; the problem was that state representatives did not understand the needs of those living within them. These were two separate informational worlds.

What follows is a speculative exercise, reflecting on what the 2021 maps tell us and how they might have helped with decision making during the pandemic. The maps were designed for the women participants, with their aesthetic and visualisations tailored to them. Had we produced maps for decision makers back then, we would have chosen other visual cues, perhaps more in line with the sorts of maps and graphics civil servants are used to.[6] Nevertheless, in the following example we can observe a number of difficulties that were experienced intensely by the women but were not addressed by policy makers (Figure 11.1).

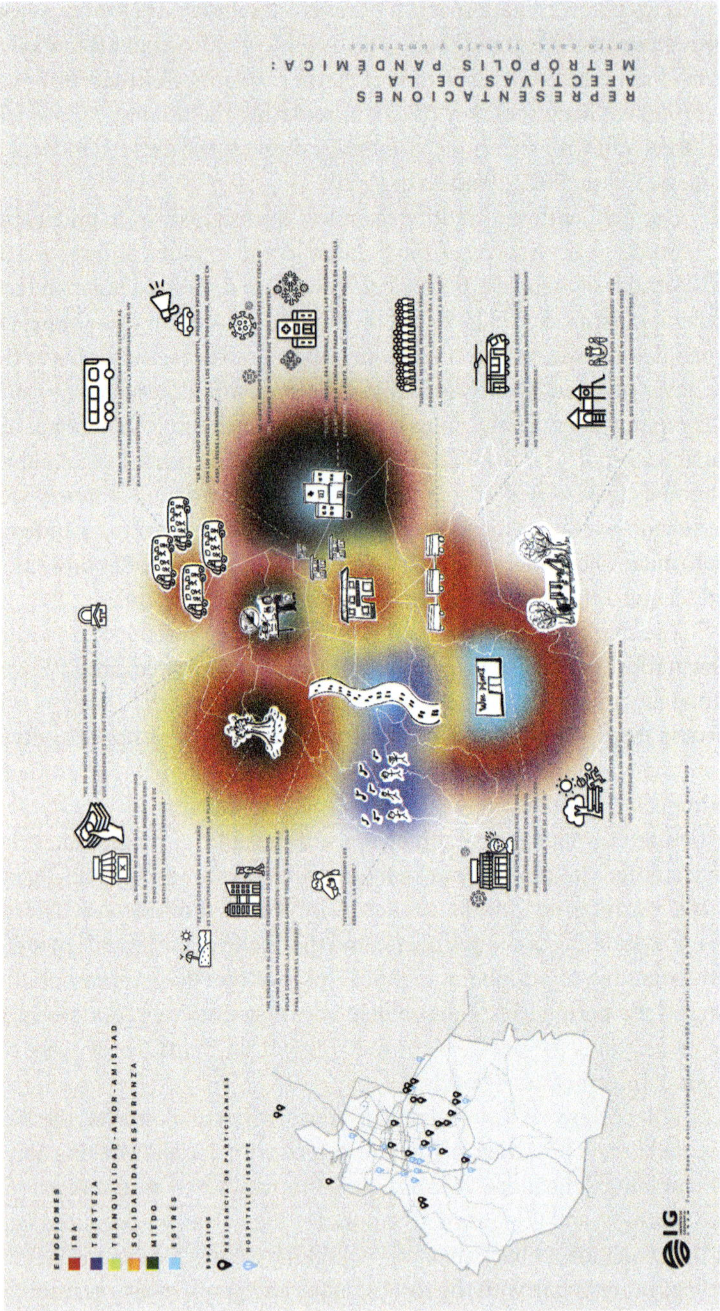

Figure 11.1 Affective representations of the pandemic metropolis, produced from a creative mapping workshop with 21 women in Mexico City, May 2021. Source: conceptualised by the research team, elaborated by Culturans.

Based on the systematisation of the testimonies and drawings collected during the workshops, this map represents some of the places and events that marked the women's pandemic peripheral realities. In the eastern peripheral borough of Iztapalapa we can see the collapsed subway line,[7] which is associated with anger (in red). We can also see how the women expressed their economic desperation as they were losing their jobs, and the moral judgement they suffered at the hands of those who could work from home, who accused them of being irresponsible for taking public transport.[8]

If we transpose this map into a format more familiar to decision makers, the difficulties mentioned by the women can be summarised using the following categories: difficulties in meeting the demands of work; accompanying their children in their online schooling; illness; living in a confined space; lack of domestic help; lack of time; bored children; economic problems; relational or family problems; technological problems; mental health; transportation; and coordination of care duties. Faced with these difficulties, the women developed various professional, relational, temporal and, above all, spatial strategies. Figure 11.2 illustrates that spatial strategies (in blue) respond in particular to relational or family difficulties, mental health, illness, bored children, online schooling, economic difficulties and limited space. The graph also shows that economic problems were mitigated to a large extent through the activation of personal networks – in other words, relational strategies (in grey).

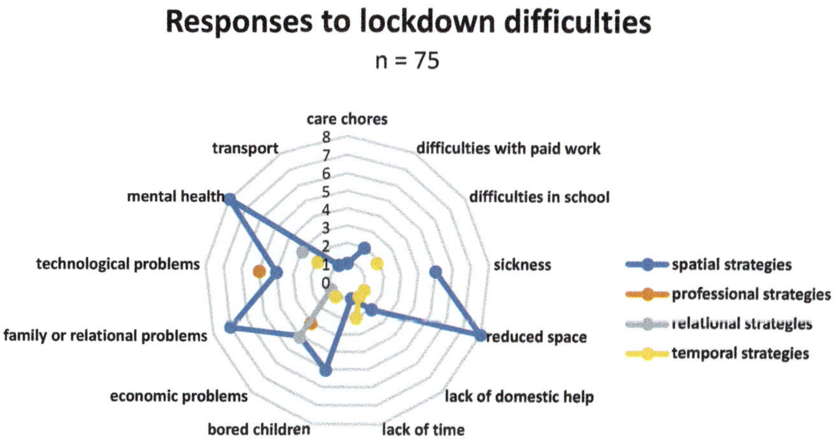

Figure 11.2 Response strategies expressed by participants in the creative mapping workshop with 21 women in Mexico City, May 2021. Source: created by the author.

The experiential data collected in the workshops reveal important factors that call for an adjustment of 'public policy shocks' in the periphery: reliance on informal work; a daily struggle to manage economic resources (no savings, no paycheques, no bank accounts); a critical lack of digital infrastructure (limited access to computers and the internet); poor and overcrowded housing conditions; and a shortage of open-air green spaces. When these factors come into play, the feasibility of lockdown measures is seriously compromised. Analysing local responses can inspire different types of 'public policy shocks' – but only if decision makers are willing to embrace partial and dynamic decision-making technologies that emerge from a different informational world.

Take, for example, the government's offer of compensation to merchants when street markets were closed. The power imbalance of street vendors' associations and corporations such as Walmart is obvious in this decision. The government's decision to close the street markets meant that food distribution would need to move to supermarkets. Many women expressed fear and resentment toward Walmart, citing experiences of harassment by security personnel and the fact that they could not enter with children, which was an impossible demand for single mothers (see Figure 11.1). In July 2020, the city launched a programme offering 3,000 pesos on electronic cards to street food vendors to compensate for their lost income. This was a one-time payment. The government distributed 2,672 cards, surpassing its initial target of 1,500. It was thus judged a success. However, there are over seventy thousand merchants on the streets of Mexico, according to most conservative estimates. Not all of them sell food, of course, but the vast majority do. One of the greatest impediments to the distribution of this form of aid was that merchants needed to be digitally registered and geo-referenced in the city databases as street vendors. The women who participated in the workshop had not even heard of the programme and were not registered in the system.

The intentions behind the government programme were good, but it really did not make a difference on the streets of the eastern periphery. In contrast, local merchants' associations helped their most vulnerable members by coordinating home-based vending, printing business cards with vendors' phone numbers and distributing them in residential areas. This street-level, non-digitalised response had a far greater impact, both for merchants and for customers. And this was evaluated very positively by the borough-level civil servant we interviewed in the periphery.

Other grassroots responses included numerous neighbourhood WhatsApp groups organising food distribution, ephemeral co-ops, and waiters who had lost their jobs offering their services as food deliverers through personal contacts and social media. We interviewed the owner of Bee Delivery, a small business in the eastern borough of Tlahuac. Founded in 2018, Bee Delivery functions like Uber Eats. It was created because commercial apps (Uber, DiDi, SinDelantal, Rappi) did not operate in the periphery. Its founder explains:

> We knew we could not compete against Rappi, against Uber Eats, at that time there was also SinDelantal, but we did say: well, we are fellow citizens, let's support the people around us, and that's where we got the idea of better ask Mrs Maria, better ask Mrs Elena, better ask Mr Rafael, Mr Jorge. (Interview 6, Bee Delivery, 18 May 2022)

These small businesses, located in the streets or within buildings, were Bee Delivery's first partners. They would call Bee Delivery when a customer required home delivery. When the pandemic hit, Bee's team grew from eight bicycle couriers to 20 delivering by motorcycle. The service enabled many small merchants to continue earning an income during the lockdown.

In 2021, Bee Delivery launched its app, which now has 62 business partners. This digitalisation succeeded because it was grounded in territorial relationships. Rather than aiming to register and geo-localise all street vendors in a government-style 'dream' database – an inherently incomplete and outdated system – the Bee Delivery app is built on the relationships that exist between neighbours. This kind of dynamic, relational digitalisation works like the street. It does not fix or produce statistics; it adapts to supply and demand, constantly shifting locations, and is founded on trust. It *composes* a possible future instead of *projecting* it. And this was possible because the app was conceptualised on the basis of territorial knowledge, rather than points on a map, curves or statistics. Making these micro-initiatives, improvisations and responses visible provides a very different lens for understanding the city's future. What would it mean if decision makers composed instead of planned?

Conclusion: speculative and planned futures from the periphery

In the face of an emergency, civil servants need to respond. I have suggested here that if their first reflex is to rely on planning instruments such as modelling and projections, emergency always requires repair. Repair as decision making in emergency situations means speculating and composing. Decision making on the basis of a composition of experiential knowledge means acting with partial knowledge, guided by empathy rather than calculations. As Koro suggests, 'methodologies that speculate ... do not verify, offer fixed solutions, or pretend to understand the other and different. Instead, they may care, connect, and create educated guesses and various scenarios of possibility building on the exploratory, imaginative, and visionary power of speculation' (Koro 2022, 139).

Participatory creative mapping workshops offer such spaces to foray into the future, as Hovorka and Peter (2021) would say – not spaces designed to prepare for a lack of hospital beds, but spaces to redefine and reconceptualise the problem. This is what the civil servants did when they trusted the experiential knowledge of territorial teams and their home care solutions. Creative and participatory mapping workshops do not produce rigid curves and projections; they activate thought and stimulate creative paths to get out of a dead end. This, however, is much more demanding for decision makers because it forces them to inhabit the pandemic peripheral reality, to see it and be affected by it, rather than analysing it through digital data. It forces them to navigate a different informational world.

As public decision making is largely based on simplified realities that overvalue digital data, how can we legitimise this kind of territorial peripheral knowledge? The first step would be to abandon the idea that only statistically representative samples are useful for decision making. Indeed, experiential knowledge provides idiosyncratic and partial information, not universal truths. Experiential knowledge is generated by listening, experimenting, observing. It focuses on fewer but deeper testimonies. This means that no modelling can be developed through equations, because the basic data are not numbers; they are words, sensations, emotions, intuitions. This would lead to decision making on the basis of speculation rather than projection, on empathy rather than calculus.

In the midst of a crisis, there is not much time for organising citizen consultations or other institutionalised forms of participation. But this is not what I am proposing here. Instead, there are two simple mechanisms that could produce more experiential and embodied information for decision making. The first is what the central city government did when hospital capacity reached its limit: they called the 'territorial team' and asked them to implement home care and fast-testing modules across the city, trusting their street-level knowledge instead of equations. Building on this positive experiment, the ADIP's director took the same approach when tasked with organising the vaccination campaign. He insisted on decentralising decision making so that each centre had enough autonomy:

> The first thing the mayor asked of all the green vests was to teach people that those of us who are vaccinating have a mystique of service [an ethics of serving] and those who get vaccinated have to feel that we are not doing them a favour, that what makes us happy is to be able to provide that service ... What we wanted was a vaccination that felt like a celebration, and not something that felt like another bureaucratic formality, which is what happened in other countries, and other states. (Interview 9, ADIP, 2 June 2022)

The second mechanism that could produce more territorial and peripheral knowledge as the basis for decision making is the organisation of decentralised focus groups. This can be done relatively quickly. In some ways, it is what our creative mapping workshop did. The great advantage of listening to targeted groups of people and tapping into their experiential knowledge through mapping tools is that it becomes easier to produce decision-making technologies that are malleable. A graph or map is much more useful to decision makers than a long report filled with interview quotes. In short, data visualisation is key to decision making. This is why mapping played such a central role during the pandemic (Boudreau forthcoming). The question is: what sorts of maps do justice to territorial peripheral realities?

Peripheral realities can only be understood if the centralised city-state informational world recognises that there is a peripheral informational world that works very differently: with different tools, sensitivities and forms of information. Decision making would be much more effective if these informational worlds were complementing one another, rather than the peripheral informational world being subordinated to the centralised one. In short, the goal is to encourage modes of state intervention that are more attuned to *caring with* (in a speculative logic) than *caring for* (in a planned logic). Deploying territorial teams, setting up a Bee Delivery app, feeling a single mother's despair when she cannot enter Walmart with her child, feeling relief as you stretch your body out of a window to get fresh air in a claustrophobic lockdown household ... These embodied practices of inhabiting the immediate future of 'public policy shocks' can be shared with policy makers in various spaces for emergency decision making. And in these spaces (be they participatory mapping workshops or focus groups), other ideas will emerge to better repair the pandemic city.

Notes

1 In addition to broadening the concept from 'assemblage' to 'world', let us be reminded, following Datta (2024), of the difference between data and information: 'Data are the basic individual items of numeric or other information, garnered through observation; but in themselves, without context, they are devoid of information. Information is that which is conveyed, and possibly amenable to analysis and interpretation, through data and the context in which the data are assembled' (Zins 2007, 479, cited in Datta 2024, 3).

2 Considered the worst industrial accident in history: half a million people were exposed to methyl isocyanate gas.

3 https://coronavirus.jhu.edu/map.html (accessed February 2025).

4 Analysing elements such as the use of the Mercator projection, the colours and the aesthetic of these cartographic representations, these authors point to the cognitive and affective reception of the maps for their readers.

5 It is important to note that Mexico City's first lockdown was initiated on 17 March 2020. Three months later, as of 15 June 2020, there was gradual reopening of commerce and subway stations. On 20 December 2020, lockdown was reinstated until 27 April 2021 – another four months of strict lockdown for the second wave of the virus. Elementary and secondary schools were closed from March 2020 to August 2021 – 15 months of enforced online schooling.

6 These maps were conceptualised by the research team. We imagined what they would look like on the basis of a qualitative systematisation of our data and then synthesising the women's drawings and words. We contracted a graphic designer to produce them using Illustrator.

7 On 4 May 2021, in the midst of one of the pandemic lockdowns, an elevated subway line fell in Mexico City, causing 26 deaths and injuring 98.

8 For more information on the methodology and the details of these maps, see Boudreau 2023.

References

Ananth, V. 2020. As Covid-19 cases rise in India, 'Covtech' based surveillance intensifies. *The Economic Times*, 30 May. Accessed February 2025. https://economictimes.indiatimes.com/news/politics-and-nation/as-covid-19-cases-rise-in-india-covtech-based-surveillance-intensifies/articleshow/74876078.cms?from=mdr.

Boudreau, J. A. 2022. City of repair: Practicing the future in Mexico City. *International Journal of Urban and Regional Research* 46(5): 851–64. https://doi.org/10.1111/1468-2427.13102.

Boudreau, J. A. 2023. Reassembling care in pandemic Mexico City. *Gender, Place & Culture* 31(11): 1625–46. https://doi.org/10.1080/0966369X.2023.2294259.

Boudreau, J. A. (forthcoming). Réparer la ville après le passage de la COVID. Cartographie comparative et collaborative à Mexico et ailleurs. *Mappemonde.*

Bryant, R. and Knight, D. M. 2019. *The Anthropology of the Future.* Cambridge: Cambridge University Press.

Dalton, M. C., Taylor, L. and Thatcher, J. 2016. Critical data studies: A dialog on data and space. *Big Data & Society* 3(1). https://doi.org/10.1177/2053951716648346.

Das, V. 1997. Sufferings, theodicies, disciplinary practices, appropriations. *International Social Science Journal* 154: 563–72. https://doi.org/10.1111/j.1468-2451.1997.tb00045.x.

Das, V., Kleinman, A., Lock, M., Ramphele, M. and Reynolds, P. (eds). 2001. *Remaking a World: Violence, social suffering, and recovery.* Berkeley, CA: University of California Press.

Datta, A. 2023. The digitalising state: Governing digitalisation-as-urbanisation in the global south. *Progress in Human Geography* 47(1): 141–59. https://doi.org/10.1177/03091325221141798.

Datta, A. 2024. The informational periphery: Territory, logistics and people in the margins of a digital age. *Asian Geographer* 41(2): 125–42. https://doi.org/10.1080/10225706.2023.2253233.

Datta, A., Aditi, A., Ghoshal, A., Thomas, A. and Mishra, Y. 2021. Apps, maps and war rooms: On the modes of existence of 'COVtech' in India. *Urban Geography* 42(1): 382–90. https://doi.org/10.1080/02723638.2020.1807165.

De Coss-Corzo, A. 2021. Patchwork: Repair labour and the logic of infrastructure adaptation in Mexico City. *Environment and Planning D: Society and space* 39(2): 237–53. https://doi.org/10.1177/0263775820938057.

Escobar, A. 2018. *Designs for the Pluriverse: Radical interdependence, autonomy, and the making of worlds*. Durham, NC: Duke University Press.

Graham, S. and Thrift, N. 2007. Out of order: Understanding repair and maintenance. *Theory, Culture & Society* 24(3): 1–25. https://doi.org/10.1177/0263276407075954.

Guma, P. K., Mwaura, M., Njagi, E. W. and Akallah, J. A. 2023. Urban way of life as survival: Navigating everyday life in a pluriversal global south. *City* 27(3–4): 275–93. https://doi.org/10.1080/13604813.2023.2214961.

Harrison, S. 2019. 'We need new stories': Trauma, storytelling, and the mapping of environmental injustice in Linda Hogan's *Solar Storms* and Standing Rock. *American Indian Quarterly* 43(1): 1–35. https://doi.org/10.5250/amerindiquar.43.1.0001.

Hovorka, D. S. and Peter, S. 2021. From other worlds: Speculative engagement through digital geographies. *Journal of the Association for Information Systems* 22(6): 1736–52. https://doi.org/10.17705/1jais.00708.

Jackson, S. J. 2014. Rethinking repair. In *Media Technologies: Essays on communication, materiality, and society*, edited by T. Gillespie P. J. Boczkowski and K. A. Foot, 221–40. Cambridge, MA: MIT Press.

Jackson, S. J. and Kang, L. 2014. Breakdown, obsolescence and reuse: HCI and the art of repair. *CHI '14: Proceedings of the SIGCHI Conference on Human Factors in Computing Systems*, 449–58. https://dx.doi.org/10.1145/2556288.2557332.

Kent, A. J. 2020. Mapping and counter-mapping COVID-19: From crisis to cartocracy. *The Cartographic Journal* 57(3): 187–95. https://doi.org/10.1080/00087041.2020.1855001.

Kitchin, R., Lauriault, T. P. and McArdle, G. 2015. Knowing and governing cities through urban indicators, city benchmarking and real-time dashboards. *Regional Studies, Regional Science* 2(1): 6–28. https://doi.org/10.1080/21681376.2014.983149.

Koro, M. 2022. Speculative experimentation in (methodological) pluriverse. *Qualitative Inquiry* 28(2): 135–42. https://doi.org/10.1177/10778004211032535.

Manning, E. 2016. *The Minor Gesture*. Durham, NC: Duke University Press.

Nancy, J-L. 1993. *Le sens du monde*. Paris: Éditions Galilée.

Pase, A., Presti, L. L., Rossetto, T. and Peterle, G. 2021. Pandemic cartographies: A conversation on mappings, imaginings and emotions. *Mobilities* 16(1): 134–53. https://doi.org/10.1080/17450101.2020.1866319.

Puig de la Bellacasa, M. 2017. *Matters of Care*. Minneapolis, MN: University of Minnesota Press.

Scott, J. C. 1998. *Seeing Like a State: How certain schemes to improve the human condition have failed*. New Haven, CT: Yale University Press.

Simone, A. and Castán Broto, V. 2022. Radical unknowability: An essay on solidarities and multiform urban life. *City* 26(5–6): 771–90. https://doi.org/10.1080/13604813.2022.2124693.

Sonn, J. W. and Lee, J. K. 2020. The smart city as time-space cartographer in COVID-19 control: The South Korean strategy and democratic control of surveillance technology. *Eurasian Geography and Economics* 61(4–5): 482–92. https://doi.org/10.1080/15387216.2020.1768423.

Ticktin, M. 2021. Building a feminist commons in the time of COVID-19 (blog). *Signs: Journal of women in culture and society*. Accessed February 2025. http://signsjournal.org/covid/ticktin.

Waisbord, S. and Segura, M. S. 2020. COVID-19 pandemic and biopolitics in Latin America. *Media Development* 3: 5–8.

Watson, C., Lupton, D. and Michael, M. 2021. The COVID digital home assemblage: Transforming the home into a work space during the crisis. *Convergence* 27(5): 1207–21. https://doi.org/10.1177/13548565211030848.

Zins, C. 2007. Conceptual approaches for defining data, information, and knowledge. *Journal of the American Society for Information Science and Technology* 58(4): 479–93. https://doi.org/10.1002/asi.20508.

12

Infrastructural injustice: the peripheralisation of urban water consumers in Lima, Peru

Fenna Imara Hoefsloot

One of the starkest examples of socio-spatial peripheralisation is the literal wall that stands between one of Lima's wealthiest suburbs and its neighbouring organically built district. In colloquial speech amongst Lima's residents, this wall is referred to as *el muro de la vergüenza* ('the wall of shame', Figure 12.1). It runs for 10 kilometres, separating the rich and poor living on two sides of the same hill. On the east side of the wall, one can see the private pools of Lima's urban rich and artificially irrigated public and private green areas. On the west side of the wall, one can see part of the expansive, newly urbanised areas within the district of San Juan de Miraflores. On the west side, many people are not connected to the water infrastructure. While this is one of the most visually accessible examples of the infrastructural injustices in Lima, many others are enmeshed in the fabric of the city or underground and out of sight.

A city characterised by what Fernández-Maldonado (2008) refers to as 'reverse urbanisation', Lima rarely follows the planned trajectory. Past urban development often exposes infrastructural plans that were never fully realised, frequently built on capitalist and modernist dreams. At times, these infrastructures and the anticipation they hold are openly visible, such as the never-entirely-constructed flyover bridges scattered along the Costa Verde, Lima's coastal road. These were designed to integrate the beach with the districts on the cliffs, but they now form a series of disconnected staircases scattered along the coast. In other instances, anticipated futures are only visible on paper, such as the zoning plans once made for the district of Pachacútec, in which the

Figure 12.1 Aerial image of Lima. The central line separating the villas on the left from the autoconstructed houses on the right is the 'wall of shame', seen from the sky. Image of Lima, Peru (12°07'21") from Google Earth Version 9.165.0.1. © Maxar Technologies. Retrieved 28 March 2021.

planners had delineated roads, parks and schools, anticipating organised urban development for the middle class. In practice, the middle class was not interested in moving to the city's far north, and so the urban poor occupied the land, rezoning it according to their own logic and incrementally building their homes, neighbourhoods and infrastructures. While this process of incremental urbanisation has seen varying degrees of success over the years, there is a general recognition that this type of urban growth is unsustainable for the future as it often results in socially segregated communities on terrains which are difficult to connect to essential infrastructure and prone to landslides (Fernández-Maldonado 2015). Nevertheless, as Fernández-Maldonado notes, a lack of public policy makes it difficult to create a viable alternative, resulting in tensions between the already existing city and its imagined future. As a result, injustices – structured and ordered through social, material and informational boundaries, generating a city of multiple peripheries in the process – are prevalent in all infrastructures in Lima.

I use the term 'infrastructural injustice' in this chapter to emphasise how the social, material and informational peripheries in Lima are not simply a misfortune or the normal order of things. Instead, they represent an injustice rooted in inequality, sustained by their structural and enduring nature, the failure of those in power to intervene, and the silent acceptance of the privileged classes (Shklar 1990). Whether in the context of injustices in the city or in the water distribution system – the focus of this chapter – we have to see these injustices in relation to the socio-technical configurations that have allowed this situation to arise and which have not (successfully) intervened to right this wrong (Moroni 2020). Researching infrastructural injustices necessitates unpacking and questioning the 'normal order' within the infrastructural system. This involves paying attention to how infrastructural injustices are created, experienced and perpetuated through everyday interactions (Rusca and Cleaver 2022), while also considering how the legacies of already existing technologies shape these injustices (Lin et al. 2015; Mattern 2021a).

In this chapter I study the processes underlying the datafication of Lima's water infrastructure. Datafication – the transformation of something, such as social activities, objects and their characteristics, or natural phenomena, into quantitative data through diverse actors, methods and technologies which allow it to be recorded, analysed and reorganised (Mayer-Schönberger and Cuckier 2013) – is frequently ascribed a key role in improving urban resource management and bringing these processes into the 'modern' age. For Sedapal (Servicio de Agua Potable y Alcantarillado de Lima – the city's metropolitan water company) and urban water consumers, there are at least three promises of modernity (Harvey and Knox 2012) that the digital infrastructure – that is, the supervisory control and data acquisition system (SCADA) – seems to make: the promises of seeing, control and economic efficiency. At first glance, these promises engage with problems that Lima's residents face. The promise of seeing relates to transparency, clarifying where water flows within the city and whom it reaches. The promise of control engages with the issue of continuity and reliability of service provision. Finally, the promise of economic efficiency speaks to the continuous critique that the water authority, as a public institution, is excessively bureaucratic.

However, while the digital infrastructure improves seeing, control and efficiency for Sedapal, this does not automatically lead to improvements in the water distribution system for all residents. As many have argued before me, datafication and digitalisation are often based on the shaky logic that more regulation and greater efficiency will also lead to a fairer distribution of urban resources (Datta 2018; Ferreira et al. 2022; Kitchin 2015). Previous research in Peru has illustrated how data-driven water governance, which focuses on quantifiable information and increasing efficiency, reduces water to its properties that can be quantified and tends to overlook labour and expertise from residents who work within the water infrastructure to overcome the service gap (Criqui 2020; Hoefsloot et al. 2020). While some elements of water are increasingly captured through legibility-making practices, these databases further fragment the conceptualisation of water and peripheralise other types of information and approaches to urban water governance. By doing so, digitalisation reproduces the unjust structures of the water distribution system through digital technologies that reinforce the framing of people who are not officially connected to water as illicit and problematic (Hoefsloot, Richter et al. 2022).

In the following sections, I will draw on Mattern's (2021a) concept of grafting and the literature on information infrastructures and critical data studies to understand how the city and its water infrastructure understand the infrastructural injustices that result in the production of the informational periphery. The understanding that information infrastructures are structuring and exclusionary, leading to infrastructural injustice, and the power of datafication in shaping the city are brought together in the concept of the informational periphery (Datta 2023; 2024). Datta (2024) explains how the informational periphery is produced through the paradoxical movements between visibilising and invisibilising. Informational peripheries are the spaces that are at once left behind and overlooked in the digitalisation of administration and the datafication of urban life, while simultaneously being targeted by the apparatus of the digital economy that sees the periphery as an opportunity for growth, as space which can be extracted from. Analysing these dynamics through grafting (Mattern 2021a), we see how these paradoxical logics and information flows are brought together and exist as a grown-together hybrid within the city. The concept of grafting explains the unity of seemingly incompatible systems of materiality, knowledge and practice. This presents the periphery not only in terms of what it lacks – planning, water, data – but as an epistemic and creative space of knowledge generation, innovation and resistance (Schmid 2018): a space where new – or ancient – logics and information are grafted onto the city.

Grafting infrastructure

In her book *A City is Not a Computer* (2021a), Mattern describes acts of urban infrastructuring as a process of grafting: the age-old horticultural technique in which one part of a plant is joined to the stem or root of another. Together, they grow into hybrids but retain distinguishable identities and origins. In my analysis of Lima's water infrastructure, I use the concept of grafting to understand how grafted infrastructures can take root and thrive, despite the economic forces pushing for uniformity, structure and artificial knowledge. Mattern (2021a) argues that through grafting, arboreal intelligences enter the conversation. By arboreal intelligence, Mattern (2021b) refers to both the metaphorical use of trees to conceptualise knowledge and learning (such as decision-making trees, disciplinary roots or random forests in machine learning) and the literal knowledge of the city embodied by nature. The latter can only be observed by paying attention to how nature responds to and interacts with changes in the human and built environment. For example, the death of trees due to heat islands or the flourishing of urban greenery during the pandemic slowdown can serve as indicators of the health, mobility and use of urban public spaces.

Although in this chapter I largely focus on digital information, following Mattern, this conceptualisation of information extends beyond numerical statistics to include knowledges, information and data that are not captured through computers, sensors or meters, but are embodied and tacit. For example, Alencastre Calderón's (2012; 2013) work brings to the fore the knowledge of marginalised urban and rural residents who are often overlooked or ignored. His research on ancestral water governance practices in the Andean highlands, or the tacit knowledge of Lima's urban dwellers regarding water management in its peri-urban areas, has been essential in bridging knowledge systems and opening up thinking about urban water governance beyond the dominant neoliberal frameworks (Alencastre Calderón 2013). By creating space for and paying attention to the many ways in which different actors, such as peripheralised people and more-than-humans, shape the city and structure infrastructures, new knowledges are grafted onto the urban.

The concept of grafting has roots in the 'infrastructural turn' (Addie et al. 2020; Amin and Thrift 2017; Graham and Marvin 2022) in geography and draws this body of literature into conversation with other strands of scholarship focusing on the datafication of the city as described in critical data studies (Heeks and Shekhar 2019; Kitchin and Lauriault 2018; Taylor and Broeders 2015). Within urban geography, the turn towards a relational ontology has resulted in an increased interest in the socio-materiality of the city and an expanding body of scholarly work providing rich analysis of the social and political nature of infrastructure (including Criqui 2020; Loftus 2007; Luque-Ayala and Marvin 2015; Niranjana 2022). This strand of urban research centres the relationships between information, bodies and the state, and how these shape everyday urban lives (Cardullo 2020; Datta 2018; Offenhuber 2017) and emphasises how infrastructure forms the support, or the essential fundament, for urban practice (Milan and Treré 2019). Enmeshed in daily life, infrastructure allows for interaction between nodes within the city. It facilitates the flow of water, electricity, people, goods, data and ideas.

However, while infrastructure provides a good entry point to conceptualise the organisation of material and social life in the city, it tends to brush over the logics underpinning these systems. Here, Mattern's concept of grafting helps us as it emphasises the different knowledges and logics that are brought together, and grow together, in the urban. Mattern extends the idea of infrastructure to the coming together of different intelligences, opening up a further understanding of the role of information in shaping the urban.

By focusing on the everyday, we move away from trying to locate power, and instead pay more attention to the situated moments wherein information infrastructure engages and enacts social relations and the (un)settling of these dynamics. By researching the mundane rather than the monumental, we can give prominence to the people, places and knowledges that are consistently excluded from dominant narratives regarding the promises that information infrastructures evoke. Additionally, doing so helps us better understand how daily struggles over resource access and distribution, visibility in data and representation in decision making are tied to ongoing struggles over social and environmental justice, identity and power (Álvarez and Coolsaet 2020; D'Ignazio and Klein 2020; Mattern 2021a; Ricaurte 2019).

Contributing to these discussions, the metaphor of grafting preserves the specificity of the urban assemblage in relation to what is being added. It sees infrastructure as changing but not replacing. Emphasising the potentialities and restrictions of legacy infrastructures, grafting helps explain how the effects of infrastructure innovation do not play out identically across cases or even within metropolitan areas. While digital technologies and 'smart systems' are said to introduce efficiency and remove human error and biases, they grow upon and within the existing institutional framework and political economy (Chambers 2020). On the other hand, grafting also helps us understand how the newly introduced infrastructure contains its own logic which is imposed on legacy systems.

Grafting can happen on a large scale, such as when a central control room is established to manage infrastructure at the metropolitan level, but it also happens through street-level, minor interventions. It is the installation of a smart water meter on decades-old pipes, or the connection of a rooftop tank as a buffer during water cuts. By paying attention to how both the government and citizens graft, grafting as a concept becomes a form of reparation, recognising the overlooked labour of people who are constructing the infrastructure and cultivating futures that envision the city's flourishing (Mattern 2021a). It can be the strategy by which the informational periphery unmakes itself, by creating alternative data infrastructures grafted onto existing structures that exclude them. Infrastructure's power, as far as it structures and is structured by its installed base, links with specific communities of practice and the norms it embodies (Star and Ruhleder 1996), is important as construction, breakdown, maintenance, repair or appropriation are moments in which the pre-existing condition, or the installed base of the infrastructure, can be negotiated (Castán Broto and Bulkeley 2015; Karasti et al. 2016). They are also the moments that illustrate how people participate in the process of grafting infrastructure and articulate an imagined future for the city (Ribes and Finholt 2009).

Therefore, this chapter aims to understand how grafting information infrastructures – understood here as a socio-technical system for generating, distributing, mobilising and contesting knowledge – plays an important role in stabilising or rearranging unjust orders within the city. I pay particular attention to the active role of residents in the process of datafication and analyse the information infrastructure implemented within the context of the wider processes of peripheralisation and infrastructural injustices.

Methodology

This chapter is part of a larger doctoral research project aimed at investigating the knowledge infrastructures of water governance in metropolitan Lima and exploring pathways for more water and data justice. The empirical material used for this chapter was collected by means of semi-structured extended interviews, focus groups and field visits between September 2019 and March 2020. Specifically, I conducted 31 expert interviews, 2 expert focus groups, 19 interviews with urban residents, and 10 in-person and 4 online focus groups with residents from three areas in Lima: José Carlos Mariátegui, Barrios Altos and Miraflores. Through two visits to the village of San Pedro de Casta, a tour of the water infrastructure in the region and conversations with community leaders in the village, I learned about the perspectives of rural residents regarding water governance in the region. In addition, interviews and a focus group with Sedapal employees and two tours at La Atarjea, Lima's main water treatment plant and Sedapal's headquarters, helped me understand the infrastructure from the perspective of Sedapal.

Together, the analysis from these five key sites revealed the diversity of physical elements, actors, procedures and information involved in water governance in Lima and San Pedro de Casta. As key fieldwork sites, they are representative of the divergent socio-economic, political and symbolic places in the city and the region. Beyond these five key sites, I visited and interviewed residents in the Villa María del Triunfo, Barranco and Comas districts of Lima and Pachacútec in Callao. The information I collected during these additional visits was important in contextualising what I had learned in the key sites and enabled me to assess the extent to which the insights gained are representative of other districts of Lima and Callao.

Combining a variety of qualitative methods, including interviews, focus groups and field visits, gave me an opportunity to highlight actors' experiences of the information and water infrastructure. With this data, I was able to analyse the processes by which urban and rural residents manage their everyday social and material worlds. To partially account for the intersectional structures that shape these experiences, I worked with and interviewed people across socio-economic classes, gender identities, ages, ethnicities and geographies.

However, as a European researcher and an outsider to the city, I do not have the lived experience that many people have regarding the enduring infrastructural injustices described in this chapter. During fieldwork I lived in the affluent neighbourhoods of Lima, where I had continuous access to water. As a short-term visitor, I did not have to worry about registering with the water company or negotiating with contractors about mapping my neighbourhood or installing water pipes. My experience of Lima is very much one of privilege. Hence, to do justice to the experiences of people on the other side of this inequality, I rely heavily on the feedback on my analysis that I received in follow-up focus groups with residents and from colleagues at Foro Ciudades para la Vida and CENCA, two civil society organisations that have been advocating for a more just city in Lima for decades.

As this research is explicitly situated in, and particular to, the context of Lima and the region and engages with diverse expressions of smart urbanism in the Global South (see, for example, Amankwaa et al. 2021; Datta 2018; Sultana 2020; Taylor and Broeders 2015), it aims to contribute to expanding the range of urban contexts within which smart urbanism is emerging. Within this research, the city is not a passive backdrop. Instead, its residents intervene, enable, resist, adapt and transform the city in various ways. Comparing and contrasting different points of view on the water and information infrastructure, placing knowledge systems side-by-side, and researching the city across geographies helps us to understand the diverse situated experiences within the city (Niranjana 2022).

Grafting digital water systems from the central control room

Every day, Sedapal's engineers capture, treat and distribute water to the city's residents and businesses. Around 24 cubic metres of water flow every second from Lima's treatment plants to its consumers (Sedapal 2022). Lima-Callao's 50 districts and 473 hydraulic sectors are served by a complex network of pipes, valves and pumps, both large and small. At points, the network still follows the trajectory of the water distribution system constructed by the Spanish colonisers in the sixteenth century (Bell 2015). Detailed information on water flow in each hydraulic sector, valve status, water pressure, and whether domestic or commercial consumers have paid their bills is continuously monitored from the central control room at La Atarjea, Lima's main water treatment plant and the head office of Sedapal. The engineers at Sedapal play a vital role in ensuring the continuous supply of water to the metropolitan area of Lima-Callao.

Yet, despite this impressively engineered system and the continuous labour involved, up until 2019, it took Sedapal an average of 10.4 years to install water and sewerage services in unplanned neighbourhoods. Nearly a million people in Lima and Callao remain unconnected to Sedapal's water distribution system (Sedapal 2022). An even larger number of residents only receive water for a few hours each day, a couple of days a week.

Public opinion suggests that Sedapal is either unwilling or unable to improve. In 2012, 2016 and 2019, residents took to the streets to protest plans to privatise water services and to express their dissatisfaction with the current water policy (*El Comercio* 2016; 2019; Jiménez 2012). More recently, in October 2021, a group of residents of San Juan de Lurigancho – Lima's most populous and poorest district – marched in protest over the lack of maintenance on water pipes in the area. The direct trigger was that residents had received an unusually high water bill, despite the water supply being cut or limited for 11 days during September and the first weeks of October (*El Comercio* 2021). However, these protests must be understood in the context of the broader sense of injustice felt by large segments of Lima's population, driven by the city's unequal water distribution. The water infrastructure is a source of widespread discontent and controversy.

Sedapal responded to the protests by emphasising its efforts to become more efficient. Through large-scale maintenance and repair projects funded by multilateral and bilateral organisations such as the World Bank and the Japan International Cooperation Agency, and the implementation of digital infrastructure, particularly geotechnologies, to improve service provision and monitor the water distribution system, Sedapal aims to respond more quickly to breakdowns and reduce leakages in the system. Digital technologies are grafted on the water infrastructure to make it sturdier, easier to manage and grow faster.

The datafication of the water infrastructure through SCADA and the installation of sensors across the distribution network have made it possible to monitor water flow and pressure remotely, without the need to be physically present at various locations. While Sedapal was previously dependent on valve operators to monitor, control and regulate specific segments of the infrastructure, the introduction of SCADA has meant that these processes can now be carried out centrally, from a single location. Through these developments in digital infrastructure and datafication, Sedapal significantly reduced the percentage of 'non-revenue water' – the primary indicator of the system's economic efficiency – from 44 per cent in 2000 to 28 per cent of the water produced in 2021 (Sedapal 2022). In short, it has made the water infrastructure more efficient and less vulnerable. This is no small feat.

Challenges for grafting in the peripheries

Nevertheless, despite improvements on the operational side of the water distribution system, datafication of the infrastructure and the city has introduced new differences between water consumers, peripheralised labour and the knowledge of urban citizens, and prioritised operational efficiency over water justice. As illustrated at the beginning of this chapter, citizens on the 'wrong side of the wall' or living in Lima's many peripheralised communities do not receive continuous drinking water, often pay a higher price per cubic metre, and do not have water meters installed in their homes to collect data for inclusion in SCADA's databases. These irregularities in the digital and water infrastructures reveal a normative logic built into the digital system.

Mattern's metaphor of grafting is useful in thinking about the relationship between the digital and water infrastructures in Lima and understanding how unjust structures in Lima's water governance are challenged or reproduced in the current information infrastructure.

Grafting emphasises how infrastructure has roots in and mediates the strengths and limitations of the installed base (Star and Ruhleder 1996). In horticulture, grafting tends to be a tricky and finicky process, and there are many reasons why a graft might go wrong – the two entities might not be fully compatible, the rootstock might not be healthy, or the grafted scion might not align properly with the stem. Similarly, the joint between Lima's water and digital infrastructures reveals similar properties to the joint of a graft. The city's digital infrastructure is not fully compatible with the structure of its water distribution system, reproduces its unjust structures, and does not align properly.

In developing digital technologies for water governance, the technocratic approach chosen by Sedapal imagines the city as a blank slate, conveniently obscuring the underlying mechanisms, historical developments and embedded ideologies. This nexus of digital technologies, material infrastructures and knowledge systems shapes the circulation of water in the city. Over the past five hundred years, engineers, citizens and policy makers have cultivated several grafted infrastructures to make the water flow (Bell 2015). The incompatibility of the installed base and the digital information infrastructure is illustrated by the assumption that water meters correspond to individual household connections, while *quintas* – traditional communal housing compounds prevalent in the city centre – and multi-storey buildings often depend on collective water connections.

Datafication of the water infrastructure both codifies how much water is received, consumed and lost and creates knowledge about people, places and issues. Digital information infrastructures provide an architecture – or protocol (Jordan 2015) – for the socio-technical relations that govern information flows. This architecture is important because it dictates the standardisation of the types of data that can be incorporated into the system, as well as the forms of interaction and organisation permitted within the information infrastructure (Jordan 2015). In other words, it is a force that excludes data and actors that do not comply with the system's structure. In doing so, information infrastructures not only function as facilitators of interactions and the flow of information but also contribute to the production of hierarchies, determining what knowledge is valuable and whose data is represented (Hoefsloot, Richter et al. 2022; Kitchin et al. 2016).

Sedapal's limited knowledge of its water pipes in the peripheries was demonstrated on 13 January 2019, when a sewerage pipe broke in the lower area of San Juan de Lurigancho, a district in Lima. It was estimated that more than 1,500 people had been affected by the sewage water that flooded their homes, stores and streets for three days (Flores 2019). This event happened without warning, but later investigations indicated that the most likely explanation for the sudden rupture was the recent construction of an elevated light rail station directly above the main sewerage pipe. The weight of the station had caused deformities in the pipe, making it vulnerable to blockage and pressure. Additionally, secondary and clandestine water and sewerage pipes had been relocated to create space for the station, affecting the inflow and outflow of water. This most likely increased the pressure in the pipe and contributed to the rupture.

In addition to illustrating the vulnerabilities inherent in the city's transport and water infrastructures, the rupture highlighted the fact that the light rail had been constructed by the Brazilian company Odebrecht – now infamous for making illegal payments to gain concessions relating to large-scale infrastructure projects throughout Latin America (Martinez Encarnación 2019) – raising questions about the quality of the construction, the role of Lima's public officials in supervising the work, and whose needs where prioritised when planning the project.

This event illustrates not only how little was understood about what was happening below ground, with the mounting pressure on the pipes until the moment of breakdown, but also how various transport, water, financial and political infrastructures – and their interrelations – were made visible through the process of 'infrastructural inversion', whereby the background becomes the foreground, making visible the embedded structures and relations that reveal the system's workings (Bowker et al. 2007; Star 1999).

In the case of Lima's digital information infrastructure for water management, delving into the system's protocol reveals how it categorises urban water consumers. The consumption and labour of those relying on clandestine connections are recorded in the water system's many databases as 'illegal', rather than as individuals who fell victim to a poorly functioning infrastructure. In effect, not fitting within the normative and calculative frameworks of what constitutes 'good' water consumption means being peripheralised in the water distribution system.

These apparent differences between water consumers are constantly reflected in the urban landscape and the narratives surrounding water consumption in the city. For example, each year during the hottest summer months, small inflatable pools start popping up on the streets. In the poorer communities of Callao, these pools are a popular summer pastime. Yet, every year, newspapers in Lima and Callao criticise these pool users for their excessive water consumption during already water-scarce months, their occupation of public space, and the potential health risks posed by mosquitoes carrying diseases like dengue and chikungunya. It is important not to downplay these issues, particularly the health risks related to stagnant, warm water in the city, but these annual debates starkly highlight how the urban poor are disproportionately surveilled and publicly criticised for their water consumption habits. In contrast, the urban rich tend to consume far larger amounts of water without facing similar public scrutiny (Ioris 2016).

This tendency to place blame on the urban poor resonates in the language of the policy makers when they say that the poor should not have chosen to build their homes on the steep hills surrounding the city, when it is said that they lack the 'education and culture' to consume water responsibly, and when geo-technologies such as the geo-radar are used to police autoconstructed water connections. These practices illustrate how the increasing use of digital technologies to mediate the relationship between Sedapal and residents favours digitally enabled participation, while disregarding the contributions of autoconstructing residents who breach the water access gap in underserved neighbourhoods (Hoefsloot et al. 2020). Likewise, knowledge that challenges the epistemological or ontological foundations on which Sedapal operates – such as experiential knowledge and ancestral approaches to water governance – are not taken up as valuable sources of information (Hoefsloot, Martinez et al. 2022).

Built from the view of Sedapal and at the scale of the metropolitan city, the digital infrastructure grafted onto the water infrastructure legitimises the choices about what is worth recording in the data (Jasanoff 2017) and, indirectly, what counts as good water management, infrastructure expansion or forms of water consumption. This point is crucial: in developing information systems such as SCADA and determining which elements of the water system are made relevant through data production, engineers often remain implicit about their ontological claims and how, in their daily work, they favour certain values, actors and ways of knowing over others.

Grafting peripheral knowledge, creating urban intelligence

Much like how a grafted tree may take on unhealthy characteristics of the rootstock, the digital infrastructure tends to reproduce the unjust structures of the water infrastructure. Sedapal has created several ICT masterplans, covering everything from internal communication to visions of a fully integrated and digitalised water system. These digital systems are said to embody multiple modes of 'smartness': its SCADA platform represents a tremendous feat, promising data-driven efficiency and aiming to control the city's erratic water flows. Through drones and geo-radars, Sedapal can map urbanisation and water pipes in greater detail than ever before. Yet, as I have discussed, outside the city's most affluent areas, relatively few of the data-driven efficiencies envisioned by the system's creators have materialised. The city of smart water does not exist yet in the peripheries. Hence, good water governance and management, if informed by holistic approaches to water justice (Zwarteveen and Boelens 2014), struggle with a fundamental information issue.

This is where Mattern (2021a) returns to nature, using it as a lens to understand how grafted infrastructures can take root and thrive, despite the economic powers pushing towards uniformity, structure and artificial knowledge. By centring the knowledges, everyday forms of grafting and the agency of both human and more-than-human residents, this approach pluralises the concept of smartness. It reveals the places they inhabit, the flows they facilitate, and the knowledge they contribute. In this section, I will highlight two initiatives that aim to bring peripheralised knowledges to the forefront. Together, the Observatorio Metropolitano de Agua (Metropolitan Water Observatory) and Lima Cómo Vamos (Lima How Are We Doing) challenge the current data-driven focus of water governance, which overwhelmingly prioritises water's utilitarian properties and the operation of the water distribution system.

As described above, it is hardly unusual for Lima's residents to get up in arms about new developments in urban water governance. The current state of the city's water governance antagonises many; Sedapal's apparent inability to address structural injustices in the water distribution system means that information and technological interventions in the system are met with a degree of distrust by citizens. Not only is the data regarding water distribution and access treated with scepticism, as it does not match the lived reality of those inhabiting the peripheries, but civil society organisations are also frustrated by their inability to access data relating to public policy. Repeated requests for insights into the annual data on the number of water connections and the quantity of water consumed under the national Transparency of Public Information Act are met with perpetual silence from Sedapal. Information about the periphery is not only incomplete, but also inaccessible.

To bridge the information gaps and incorporate logics and knowledge from the margins into the discussion about water governance, several research and civil society organisations have developed alternative information infrastructures. In late 2022, the Observatorio Metropolitano de Agua was launched. Using crowd-sourced data to map water injustice and generate data on water access, use, reliability and cost, the Observatorio's goal is that such visualisations will enhance our knowledge of water circulation and where it does not reach, and support efforts and actors that are working towards reducing infrastructural injustice (Hoefsloot, Jiménez et al. 2022). The Observatorio seeks to accommodate diverse forms of knowledge about water, engaging communities across various socio-economic backgrounds. However, tensions have emerged between the institutional preference for quantitative data and the need to include non-quantifiable understandings of water. This underscores the challenges of decolonial efforts within existing knowledge systems (Hoefsloot, Martinez et al. 2022), as well as the difficulty of translating tacit or indigenous knowledge into measurement criteria, graphs and actionable knowledge for policy makers.

In a similar vein, the data non-profit Lima Cómo Vamos publishes regular reports that present data on the city's demographics, urban growth, economic development and the distribution of resources – including water (Lima Cómo Vamos 2022). Collected through their own means, this data serves as a supplement to the government statistics, providing data-based arguments to hold the government to account. These initiatives point towards the possibility of creating information infrastructures and data from the peripheries and working towards infrastructural justice. However, currently, they still operate on the margins of place, flows and knowledge systems.

These projects recognise the city's water infrastructure and the systems developed by peripheralised citizens as a rich ecology of knowledges, embodying the expertise of residents who, in response to the rhythms of capital – the growing city, the construction of large-scale infrastructures and the push for efficiency – have created and developed systems that are vital to their own flourishing and well-being. What these examples show is that deploying information technologies not for the sake of 'efficiency' and economic convenience, but for justice and equity, requires tremendous investment and a commitment to reconfiguring the infrastructures that have historically reproduced injustices.

Thinking about digital tools that approach water outside a productionist or utilitarian frame allows us to focus on the web of relations and materials involved in its circulation. It helps us appreciate the various ways of knowing the city and the diverse forms of intelligence and data that can, and should, inform that work. Nevertheless, at present, there is little indication of potential change. Currently built and managed by technocrats in collaboration with private corporations with little transparency and public accountability, the digital technologies available are deployed uncritically to enhance optimisation and police water consumers, rather than recognising the rich body of knowledge and intelligence about water and water infrastructures that exists within the urban population.

Towards infrastructural justice

Viewing the development of the city through the lens of grafting presents information infrastructures embedded in, and structuring, the broader urban fabric. The city is not a tabula rasa, and information infrastructures are not built on a blank slate. This research shows how the current digitalisation of urban water governance takes shape along conventional lines: it differentiates between water consumers, leads to the further peripheralisation of the non-digital citizen, and overlooks the many ways in which urban residents contribute to water management and governance. It follows that this path of digitalisation will not solve the structural injustices in Lima's water distribution system. Instead, it produces an informational periphery, where data is fragmented and generally serves to police non-metered forms of water consumption, rather than providing better public services.

This is exemplified by Sedapal, whose managerial approach embraces the distinct imaginary of a manageable city, where the control room is the central node through which all knowledge flows, and water is a commodified resource that can be governed in an economic realm. The SCADA system embodies this vision. In contrast, the city's self-building and auto-managing citizens embody neither of these ideals and consequently face decades of peripheralisation, the deconstruction of their water systems, and silencing. Normative frameworks embedded in the information infrastructure – such as the hierarchical differentiation of water consumers or the conceptualisation of water as a market resource rather than a commons, backed up by sanctions against clandestine water systems and indicators such as non-revenue water – restrict opportunities to restructure and rethink the infrastructure from the perspective of peripheralised consumers (Hoefsloot et al. 2020).

Taking grafting as a conceptual departure point, this chapter has brought together various perspectives to narrate the story of datafication in Lima's water distribution system, the ways in which residents contribute through labour and information from the peripheries, and how we can imagine more just routes that contribute to infrastructural justice. By looking at the governance of water in Lima, we see how peripheralisation is produced through place – the dynamic and heterogeneous spaces on the fringes of urban sprawl – flow – the distribution of resources, labour or services – and information (Datta 2024). The informational periphery occupies the space which is marginalised on all three counts. This understanding of the informational periphery highlights the connectedness of the spatial with the social, and the material with the anticipated.

For the communities awaiting water connections, or the people taking to the streets to protest, the urgency to expand the distribution system and provide adequate public water infrastructure is related to the desire for social and urban transformation. Both the protests and the bottom-up information infrastructures that have emerged stem from a shared recognition across districts and socio-economic classes that water access is a necessary and fundamental good. And, despite the context of discontent and prolonged struggle, Lima's urban and rural residents ultimately cultivate hope through their strategies to expand, diversify, restructure, maintain or resist the water and digital infrastructures.

But a different approach is needed to facilitate change of this kind. Aligned with earlier writings on this topic (Criqui 2020) and acknowledging Lima's legacy of incremental urbanisation and autoconstruction (Fernández-Maldonado 2015; Turner 1968), I argue that we need information infrastructures that embrace the pragmatic and decentralised approaches of Lima's peripheral citizens in constructing and managing neighbourhood water distribution systems, with justice as the departure point. This can be done top-down, from the perspective of the state, or bottom-up, driven by civil society. In both cases, it is important that the information systems used to govern the 'smart city' are built to better address human needs.

Acknowledgements

This chapter is based on research conducted as part of a PhD thesis entitled 'Knowledge Infrastructures for Just Urban Futures: A case of water governance in Lima, Peru'. I thank Karin Pfeffer, Javier Martinez and Christine Richter for their support and feedback on earlier versions of the text. Most importantly, I am grateful and indebted to the communities in Lima for generously sharing their knowledge, time and energy.

References

Addie, J. P. D., Glass, M. R. and Nelles, J. 2020. Regionalizing the infrastructure turn: A research agenda. *Regional Studies, Regional Science* 7(1): 10–26. https://doi.org/10.1080/21681376.2019.1701543.

Alencastre Calderón, A. 2012. Las amunas. Siembra y cosecha del agua. *LEISA Revista de Agroecología* 28(1): 36. Accessed February 2025. https://leisa-al.org/web/wp-content/uploads/vol28n1_leisa.pdf.

Alencastre Calderón, A. 2013. Hacia un balance de la gestión del agua: ¿Explotación de un recurso natural o gestión sostenible de un bien común? In *Perú Hoy: Susurros desde Babel*, edited by E. Toche Medrano, 327–44. Lima: DESCO.

Álvarez, L. and Coolsaet, B. 2020. Decolonizing environmental justice studies: A Latin American perspective. *Capitalism, Nature, Socialism* 31(2): 50–69. https://doi.org/10.1080/10455752.2018.1558272.

Amankwaa, G., Heeks, R. and Browne, A. L. 2021. Digital innovations and water services in cities of the Global South: A systematic literature review. *Water Alternatives* 14(2): 619–44. Accessed February 2025. https://www.water-alternatives.org/index.php/alldoc/articles/vol14/v14issue2/637-a14-2-15/file.

Amin, A. and Thrift, N. 2017. *Seeing Like a City*. Cambridge: Polity Press.

Bell, M. G. 2015. Historical political ecology of water: Access to municipal drinking water in colonial Lima, Peru (1578–1700). *Professional Geographer* 67(4): 504–26. https://doi.org/10.1080/00330124.2015.1062700.

Bowker, G. C., Baker, K. S., Millerand, F. and Ribes, D. 2007. Towards information infrastructure studies: Ways of knowing in a networked environment. In *International Handbook of Internet Research*, edited by J. Hunsinger, L. Klastrup and M. Allen, 97–117. Dordrecht: Springer.

Cardullo, P. 2020. *Citizens in the 'Smart City': Participation, co-production, governance*. Abingdon: Routledge.

Castán Broto, V. and Bulkeley, H. 2015. Maintaining experiments and the material agency of the urban. In *Infrastructural Lives: Urban infrastructure in context*, edited by S. Graham and C. McFarlane, 199–218. Abingdon: Routledge.

Chambers, T. 2020. 'Lean on me': Sifarish, mediation & the digitisation of state bureaucracies in India. *Ethnography* 26(1): 84–105. https://doi.org/10.1177/1466138120940755.

El Comercio. 2016. Sedapal: Así fue marcha en vía expresa contra inversión privada. 18 February. (Source unavailable.)

El Comercio. 2019. Marcha por el agua: Movilización contra privatización de Sedapal intenta llegar a San Isidro. 26 September. Accessed February 2025. https://elcomercio.pe/lima/marcha-por-el-agua-movilizacion-contra-privatizacion-de-sedapal-intenta-llegar-a-ministerio-de-vivienda-en-san-isidro-noticia.

El Comercio. 2021. SJL: La marcha contra Sedapal que vecinos realizaron por cobros en *último* recibo de agua y fisuras halladas en tuberías. 18 October. Accessed February 2025. https://elcomercio.pe/lima/san-juan-de-lurigancho-la-marcha-contra-sedapal-que-un-grupo-de-vecinos-realizo-por-cobros-en-ultimo-recibo-de-agua-y-fisuras-halladas-en-tuberias-fotos-nndc-noticia.

Criqui, L. 2020. Sociotechnical alternatives and controversies in extending water and sanitation networks in Lima, Peru. *Water Alternatives* 13(1): 160–81. Accessed February 2025. https://www.water-alternatives.org/index.php/alldoc/articles/vol13/v13issue1/569-a13-1-8/file.

Datta, A. 2018. The digital turn in postcolonial urbanism: Smart citizenship in the making of India's 100 smart cities. *Transactions of the Institute of British Geographers* 43(3): 405–19. https://doi.org/10.1111/tran.12225.

Datta, A. 2023. The digitalising state: Governing digitalisation-as-urbanisation in the global south. *Progress in Human Geography* 47(1): 141–59. https://doi.org/10.1177/03091325221141798.

Datta, A. 2024. The informational periphery: Territory, logistics and people in the margins of a digital age. *Asian Geographer* 41(2): 125–42. https://doi.org/10.1080/10225706.2023.2253233.

D'Ignazio, C. and Klein, L. F. 2020. *Data Feminism*. Cambridge, MA: MIT Press.

Fernández-Maldonado, A. M. 2008. Expanding networks for the urban poor: Water and telecommunications services in Lima, Peru. *Geoforum* 39(6): 1884–96. https://doi.org/10.1016/J.GEOFORUM.2007.11.007.

Fernández-Maldonado, A. M. 2015. Las barriadas de Lima como estímulo a la reflexión urbana sobre la vivienda: Revisitando a Turner y de Soto. *Revista de Estudios sobre Vivienda (WASI)* 2(3): 7–24. Accessed February 2025. https://revistas.uni.edu.pe/index.php/wasi/article/view/1752/2077.

Ferreira, A., Oliveira, F. P. and Von Schönfeld, K. C. 2022. Planning cities beyond digital colonization? Insights from the periphery. *Land Use Policy* 114. https://doi.org/10.1016/j.landusepol.2022.105988.

Flores, H. 2019. *Informe de Emergencia N° 172 – 25 Feb 2019: Aniego en el distrito de San Juan de Lurigancho – Lima (Informe N°30)*. Lima: Centro de Operaciones de Emergencia Nacional.

Graham, S. and Marvin, S. 2022. Splintering urbanism at 20 and the 'infrastructural turn'. *Journal of Urban Technology* 29(1): 169–75. https://doi.org/10.1080/10630732.2021.2005934.

Harvey, P. and Knox, H. 2012. The enchantments of infrastructure. *Mobilities* 7(4): 521–36. https://doi.org/10.1080/17450101.2012.718935.

Heeks, R. and Shekhar, S. 2019. Datafication, development and marginalised urban communities: An applied data justice framework. *Information Communication and Society* 22(7): 992–1011. https://doi.org/10.1080/1369118X.2019.1599039.

Hoefsloot, F. I., Martínez, J., Richter, C. and Pfeffer, K. 2020. Expert-amateurs and smart citizens: How digitalization reconfigures Lima's water infrastructure. *Urban Planning* 5(4): 312–23. https://doi.org/10.17645/up.v5i4.3453.

Hoefsloot, F. I., Jiménez, A., Sara, L. M., Flores, L. E., Martinez, J. and Pfeffer, K. 2022. The Observatorio Metropolitano de Agua para Lima-Callao: A digital platform for water and data justice. In *Freedom and Social Inclusion in a Connected World*, edited by Y. Zheng, P. Abbott and J. A. Robles-Flores. Cham: Springer.

Hoefsloot, F. I., Martínez, J. and Pfeffer, K. 2022. An emerging knowledge system for future water governance: Sowing water for Lima. *Territory, Politics, Governance* 12(6): 825–45. https://doi.org/10.1080/21622671.2021.2023365.

Hoefsloot, F. I., Richter, C., Martínez, J. and Pfeffer, K. 2022. The datafication of water infrastructure and its implications for (il)legible water consumers. *Urban Geography* 44(4): 729–51. https://doi.org/10.1080/02723638.2021.2019499.

Ioris, A. A. R. 2016. Water scarcity and the exclusionary city: The struggle for water justice in Lima, Peru. *Water International* 41(1): 125–39. https://doi.org/10.1080/02508060.2016.1124515.

Jasanoff, S. 2017. Virtual, visible, and actionable: Data assemblages and the sightlines of justice. *Big Data & Society* 4(2). https://doi.org/10.1177/2053951717724477.

Jiménez, B. 2012. La 'marcha del agua' en Perú divide a los partidarios del presidente Humala. *El Mundo*, 11 February. Accessed February 2025. https://www.elmundo.es/america/2012/02/11/noticias/1328932919.html.

Jordan, T. 2015. *Information Politics*. London: Pluto Press.

Karasti, H., Millerand, F., Hine, C. M. and Bowker, G. C. 2016. Knowledge infrastructures: Part I. *Science and Technology Studies* 29(4): 2–12. https://doi.org/10.23987/sts.55406.

Kitchin, R. 2015. Making sense of smart cities: Addressing present shortcomings. *Cambridge Journal of Regions, Economy and Society* 8: 131–6. https://doi.org/10.1093/cjres/rsu027.

Kitchin, R. and Lauriault, T. P. 2018. Toward critical data studies: Charting and unpacking data assemblages and their work. In *Thinking Big Data in Geography: New regimes, new research*, edited by J. Thatcher, J Eckert and A. Shears, 3–20. Lincoln, NE: University of Nebraska Press.

Kitchin, R., Maalsen, S. and McArdle, G. 2016. The praxis and politics of building urban dashboards. *Geoforum* 77: 93–101. https://doi.org/10.1016/J.geoforum.2016.10.006.

Lima Cómo Vamos. 2022. *¿Cómo vamos en Lima y Callao? Reporte urbano de indicadores de calidad de vida 2021*. Accessed February 2025. https://www.limacomovamos.org/wp-content/uploads/2022/11/ReporteIndicadoresLCV2021.pdf.

Lin, Q., Kalantari, M., Rajabifard, A. and Li, J. 2015. A path dependence perspective on the Chinese cadastral system. *Land Use Policy* 45: 8–17. https://doi.org/10.1016/j.landusepol.2015.01.017.

Loftus, A. 2007. Working the socio-natural relations of the urban waterscape in South Africa. *International Journal of Urban and Regional Research* 31(1): 41–59. https://doi.org/10.1111/j.1468-2427.2007.00708.x.

Luque-Ayala, A. and Marvin, S. 2015. Developing a critical understanding of smart urbanism? *Critical Commentary Urban Studies* 52(12): 2105–16. https://doi.org/10.1177/0042098015577319.

Martinez Encarnación, K. 2019. Dinero, poder y política: Financiamiento electoral como clave en la influencia de Odebrecht en Perú y México. *Politai: Revista de ciencia política* 10(18): 128–68.

Accessed February 2025. https://www.researchgate.net/publication/336402456_Dinero_poder_y_politica_financiamiento_electoral_como_clave_en_la_influencia_de_Odebrecht_en_Peru_y_Mexico.

Mattern, S. 2021a. *A City is Not a Computer: Other urban intelligences*. Princeton, NJ: Princeton University Press.

Mattern, S. 2021b. Tree thinking. *Places Journal*, September. https://doi.org/10.22269/210921. Accessed February 2025. https://placesjournal.org/article/tree-thinking.

Mayer-Schönberger, V. and Cuckier, K. 2013. *Big Data: A revolution that will transform how we live, work, and think*. London: John Murray.

Milan, S. and Treré, E. 2019. Big data from the South(s): Beyond data universalism. *Television and New Media* 20(4): 319–35. https://doi.org/10.1177/1527476419837739.

Moroni, S. 2020. The just city. Three background issues: Institutional justice and spatial justice, social justice and distributive justice, concept of justice and conceptions of justice. *Planning Theory* 19(3): 251–67. https://doi.org/10.1177/1473095219877670.

Niranjana R. 2022. An experiment with the minor geographies of major cities: Infrastructural relations among the fragments. *Urban Studies* 59(8): 1556–74. https://doi.org/10.1177/00420980221084260.

Offenhuber, D. 2017. *Waste is Information: Infrastructure legibility and governance*. Cambridge, MA: MIT Press.

Ribes, D. and Finholt, T. A. 2009. The long now of technology infrastructure: Articulating tensions in development. *Journal of the Association for Information Systems* 10(5): 375–98. https://doi.org/10.17705/1jais.00199.

Ricaurte, P. 2019. Data epistemologies, the coloniality of power, and resistance. *Television & New Media* 20(4): 350–65. https://doi.org/10.1177/1527476419831640.

Rusca, M. and Cleaver, F. 2022. Unpacking everyday urbanism: Practices and the making of (un)even urban waterscapes. *WIREs Water* 9(2). https://doi.org/10.1002/wat2.1581.

Schmid, C. 2018. Journeys through planetary urbanization: Decentering perspectives on the urban. *Environment and Planning D: Society and space* 36(3): 591–610. https://doi.org/10.1177/0263775818765476.

Sedapal. 2022. *Memoria Anual 2021*. Lima: SEDAPAL.

Shklar, J. 1990. *The Faces of Injustice*. New Haven, CT: Yale University Press.

Star, S. L. 1999. The ethnography of infrastructure. *American Behavioral Scientist* 43(3): 377–91. https://doi.org/10.1177/00027649921955326.

Star, S. L. and Ruhleder, K. 1996. Steps toward an ecology of infrastructure: Design and access for large information spaces. *Information Systems Research* 7(1): 111–34. https://doi.org/10.1287/isre.7.1.111.

Sultana, F. 2020. Embodied intersectionalities of urban citizenship: Water, infrastructure, and gender in the Global South. *Annals of the American Association of Geographers* 110(5): 1407–24. https://doi.org/10.1080/24694452.2020.1715193.

Taylor, L. and Broeders, D. 2015. In the name of development: Power, profit and the datafication of the Global South. *Geoforum* 64: 229–37. https://doi.org/10.1016/j.geoforum.2015.07.002.

Turner, J. C. 1968. Housing priorities, settlement patterns, and urban development in modernizing countries. *Journal of the American Planning Association* 34(6): 354–63. https://doi.org/10.1080/01944366808977562.

Zwarteveen, M. Z. and Boelens, R. 2014. Defining, researching and struggling for water justice: Some conceptual building blocks for research and action. *Water International* 39(2): 143–58. https://doi.org/10.1080/02508060.2014.891168.

13

Conclusions: towards global informational peripheries

Fenna Imara Hoefsloot and Ayona Datta

In 2021, the plans of tech giant Meta to build a new 'hyperscale' data centre in Zeewolde, a small town in the Netherlands, caused a stir in the community and led to a national conversation about decision making regarding the construction of data centres in the country. Operating under the name 'Polder Networks', Meta had been in the process of buying 166 hectares of brownfield land and, following active lobbying by the Minister of Economic Affairs and Environment and the Dutch Data Center Association (a lobby group representing, amongst others, Google), it was on the verge of securing approval to build the country's largest and most advanced data centre. The facility was estimated to consume twice as much energy as the entire city of Amsterdam (Stikker and van Eeden 2021).

The sale and construction were eventually cancelled after sustained protests from the community, local elections and a loss of political will on the part of the national government as the negotiations continued. As more information came to light, it became clear that the social costs and benefits did not paint a promising picture for the community of Zeewolde. While the data centre would take up physical space, put pressure on the electricity network and potentially lead to an urban heat island, there were few benefits. Job creation would be minimal, there was no clear plan to offset the carbon emissions from the data centre, and its proximity to a natural park raised concerns regarding the impact of increased traffic and nitrogen pollution on the environment.

One of the most critical issues revealed by the process was the stark lack of transparency and communication regarding the decision making behind bringing such digital developments to the region. The concerted lobbying efforts of both industry and the national government had put pressure on the local government to make space for the data centre, while they were poorly informed about the social and environmental costs, as well as the limited benefits it would bring. The establishment of large-scale data centres, or data campuses as they are increasingly referred to, has significant impacts on local residents, who are often not considered legitimate stakeholders in making decisions or profits. Ultimately, this case raised calls for more deliberation, inclusive decision making, and a national data centre policy that considers local impacts.

These dynamics are far from unique to the Netherlands. Across Europe, the UK and the USA, tech companies such as Meta, Google and Amazon are encroaching on peripheral lands, near strategic locations such as ports, military hubs or internet transformer centres, to create the material infrastructures on which their digital – and increasingly AI – empires are to be built (Bridges 2024; Brodie 2023; Vonderau 2019). These companies may buy land and properties directly, or they may operate through specialist development companies that target customers like Amazon and Meta as potential clients, leading to speculative land markets. These developments occur under banal names that do not point to the tech companies involved, veiling the true purpose and actors behind them. As illustrated in the example of the Netherlands, governments and developers often work together in speculating on the potential of land.

Discussing the construction of data centres in Europe and the USA may seem like a lateral step after discussing the production, territorialisation and experience of informational peripheries in the Global South. However, data centres are part of the extended nodes of informational peripheries in the Global South that are produced from geopolitical elsewheres. The issues raised by the proposed Zeewolde data centre – the pooling of state and non-state actors in the production of the digital infrastructure, the material and territorial costs of digitalisation, and the resistance to these developments from the peripheries – resonate with the themes of this book. And, by extension, we contend that the chapters in this volume help us understand how this is not just a function of the land market, political ambitions and logistics but also the production and unfolding dynamics of informational peripheries everywhere.

Seeing from 'Southern' informational peripheries

The chapters in this volume have analysed a diversity of informational peripheries in the Global South. Drawing on empirical work from Cape Town, Nairobi, Mumbai, Mexico City, Guadalajara, Lima, Guiyang, Delhi and Ahmedabad, we show how the informational periphery manifests itself in unexpected, paradoxical and diffuse ways, speaking to the diversity of materialities, governance cultures and societies from which they emerge and develop. These chapters illustrate the contentious relationships between information technologies and the materialities of paper, land and labour. They also highlight the material manifestation of digital economies evident in the wires, data centres and warehouses across strategic sites in the Global South. Datta's illustration of the informational periphery through logistics spaces in Bhiwandi (Chapter 6), Chang and Dai's analysis of the ties between information infrastructures and sustainability narratives in Guiyang (Chapter 9), and Boudreau's examination of the use of information, graphs and embodied knowledge in managing the COVID-19 pandemic in Mexico City (Chapter 11) all illustrate how the Global South is intimately linked to – and shaped by – global informational flows. Taken together, they highlight how a 'digital turn in postcolonial urbanism' (Datta 2018) is also geographically uneven, simultaneously transforming people, land and territory and creating new articulations of informational peripheries elsewhere.

In this concluding chapter, we explore what informational peripheries in the Global South can offer to urban theory at large and how they might provide an entry point for thinking about the processes of urbanisation in an AI-mediated urban future. As Halvorsen argues, the value of learning from the South, in his case Latin America, 'appears self-evident and requires little justification. Yet the form and distribution of this value is unclear and demands greater scrutiny' (Halvorsen 2018, 11). 'Seeing from the South' can inform our conceptualisation, theoretical engagement, methodologies and research agendas on peripheries, smart cities and platform urbanism globally. Seeing from the South refers to far more than the geographical location of the case studies we have included in this volume. It refers to engagement with the most marginalised, and a critical investigation of the systems that maintain structural inequalities, such as between the North and South, the landowning class and informal occupiers, global IT corporations and gig labourers. From this perspective, the South is an epistemic place (de Sousa Santos 2016; Halvorsen 2018) which can be deployed to further our investigations into issues such as labour (Datta and Muthama 2024; Muldoon et al. 2024; Woodman and Cook 2019), surveillance (Datta 2020; Edwards et al. 2018; Zuboff 2023), land and resource capture (Branch 2017; Caranto Morford and Ansloos 2021), or the politics of (dis)connection (Couldry and Mejias 2019; Tsing 2005). Specifically, seeing informational peripheries from the South offers a heuristic for understanding the social and political impacts of digitalisation on people. It transcends the binaries of the West and the rest – proximity and distance, core and periphery, cloud and earth, past and future – to extend the idea of marginality and exclusion in a digital age as, crucially, one of informational distance from those exerting power at various scales.

We argue, therefore, that instead of pigeonholing these chapters as 'Southern', we need to see the South as a site for the production of theory around informational peripheries, which can help us document the paradoxical nature of peripheralisation and marginality in a global digital age and witness the digital and geographical networks of informational peripheries globally. This follows in the footsteps of many scholars from both the North and the South who have been calling for a more dynamic exchange of ideas as 'travelling theory' (Robinson and Parnell 2011), which acknowledges contributions from the South, or 'many elsewheres' (Robinson 2022). Seeing informational peripheries as travelling across a 'world of cities' (Robinson 2011) and regions acknowledges the circularity and plurality of knowledge systems (Hoefsloot et al. 2024) across the Global North and South that a digital world makes possible.

Following this tradition, we extend the conceptualisation of informational peripheries in two ways. First, in line with Eubanks' (2018) work on automating inequality in the US social welfare system, we argue that the scholarship in this book on digitalisation and automation in the Global South shows how these processes are as much entailed in detailed operations of bureaucracy and implementation of policy as they are outcomes of external 'rupture' through rapid technological developments led by big tech. As illustrated at the start of this chapter, the development of digital systems for the state is not independent of the global digital economy, yet the role of the state in transforming metropolitan urbanisation might need more theoretical work on a global scale.

Second, we propose that, in a moment when many of us are taken aback by the spectacle of generative AI – the digitalisation of urban life has evolved way beyond the smart city, encompassing robotics, autonomous technologies, digital twins, city brains and a staggering diversity of possible cyber-enhanced futures – the research in this book grounds these developments in their historical development from paper, to digital, to algorithmic systems, offering a conceptual starting point to locate their effects. This is distinct from discussions on AI urbanism (Cugurullo et al. 2023), which tend to focus on hyperconnectivity, autonomous vehicles, robotics, drones or the impact of AI-supported tools for the automation of routine labour and processes (Batty 2018; Herath and Mittal 2022).

Bring the state back in

Returning to Zeewolde and the decision making around the construction of the hyperscale data centre, one issue that stands out is the dynamics between the different scales of government in granting permission to build. These scalar politics point to the multiplicity of the 'state', not only in terms of administrative power but also regarding policy and the external powers it responds to. The state was present in both the foreground and background, acting in both visible and invisible ways. Hence, in this first extension, we argue that we need to bring the state back into our thinking and analysis of digitalisation.

In the Global North, debates about the role of digitalisation and datafication have largely focused on the role of corporations, specifically big-tech and algorithmic networks, in mass surveillance capitalism at the urban scale (Söderström and Datta 2023; Zuboff 2015). Although the role of corporate actors in the digitalisation of public life tends to be discussed in relation to the state, either through the conceptualisation of state-corporate networks such as public–private partnerships, smart city ecologies or the triple helix (Mello Rose et al. 2022), or by looking into the ways that corporate actors themselves start performing the tasks of the state (Zuboff 2015), these conceptualisations tend to be skewed towards the position of companies and their pursuit of growth. However, as Brodie argues in his work on the development of global internet infrastructures in Ireland's peripheries, 'digital growth looks different across global geographies, and responds to lumpy infrastructural affordances and environmental antagonisms in local jurisdictions' (Brodie 2023, 4).

To capture the dynamics between actors, be they state, corporate, civil society or knowledge institutions, we need to link these networks to the rhythms of place-based urban developments (Barns 2016) and trans-scalar politics. Like the case of Zeewolde, where the ministry, local government and civil society groups all negotiated in the production and resistance of the informational periphery, this book has shown that 'peripheralisation' is a process (rather than just an outcome) occurring through the circulation of information through certain spaces, institutions and communities at the cost of bypassing others. The chapters in this volume illustrate how the informational periphery is produced from the remnants of colonial bureaucracy that enables files, folders and ledgers to capture and circulate information and produce knowledge of the state, the labour of casualised workers within bureaucratic offices that support its digitisation, and the street-level informational infrastructures within communities and neighbourhoods. The state is central to the production, territorialisation and experience of the informational periphery.

Hence, we want to put on the agenda the importance of researching the implications of technological development for the workings of local, regional and national government and its fundamental role in governance and planning over land, resources and people. Being sensitive to how these are also connected to broader trends and normative pressures of the political economy of digital knowledge production requires that we engage simultaneously with state theory and theories of grassroots citizenship when we talk about urban digitalisation. Looking at the materiality, geography and social life of the digital rather than its disembodied form – the 'cloud' or the 'internet' – shows how digitalisation is happening in conjunction with settler dynamics and land dispossession in the peripheries, in the South as well as in the North (Vonderau 2019). This is particularly evident in the wealth of critical research emerging from scholars of Northern cities who draw on decolonial and feminist frameworks and language to understand and describe the interplay between structural deprivation, state welfare schemes and digitalisation (Benjamin 2019; Eubanks 2018; Mattern 2021). These works show how those living in the peripheries of the central powers are continuously forced to navigate the fluidity of technological and urban change, the uncertainty of the future, and the fallibility of the state to care. This peripheralisation, as Gururani (2024) has argued, is a permanently incomplete process that has a *longue durée* of manifestation across colonial, postcolonial and neoliberal informational governance systems and informational flows. Peripheralisation shapes the enduring relationship between land, territory, state, and a multitude of global informational spaces. As a process, therefore, it offers us ways to connect the diversity of inequalities and exclusions at different scales and spaces as forms of 'expulsion' (Sassen 2014) from the core of knowledge.

AI peripheries

The second extension is to ground our thinking about informational peripheries in the historical development of incomplete and territorialised information infrastructures that preceded the emergence of AI technologies. While the rapid integration of algorithmic, automated and intelligent technologies into our daily lives has been impressive, the techno-utopian fetishisation around the limitless possibilities of digital technologies, as well as the worries about their impact, are far from new. The legitimate concerns regarding biases and representation in the data and knowledge production with generative AI, accessibility of the technology, and the small number of companies that develop and control these systems all echo the concerns raised by scholars of digital governance and geography regarding computational technologies, dashboards and smart city-era surveillance systems (Amoore 2018; Eubanks 2018; Kitchin and Dodge 2017; Maalsen et al. 2023; Taylor 2017), illustrating how the rise of big data and algorithmic power is situated in the long waves of informational transformation that reinforce or limit their impact.

In the Global North, discussions on the proliferation of generative AI and biometric technologies – often used in predictive policing and by private security actors – focus on the ethical practices and regulations required to protect human rights and democratic institutions. Calvo (2020), in outlining the diverse issues associated with the hyperconnectivity, datafication and algorithmication of the city, raises concerns about the ethical implications of these digital technologies. They include issues related to privacy, inclusion, biases in algorithmic decision making, the misuse of data, and the dissolution of accountability and responsibility in urban governance, driven by the opacity of technological implementation and market-driven logics. While these issues are certainly more muted in the Global South due to the legacy of more rudimentary and frugal forms of informational infrastructures coexisting alongside the digital, it is not the case that they are absent altogether.

For example, AI-supported automated systems are increasingly being integrated into governance in Global South countries. In India, recent laws and policies on biometric identification and data privacy identify highlight the growing use of AI for state surveillance and expanding the state's power to govern at a distance (Bhatia and Bhabha 2017; Dattani 2020; Sarkar 2014). Since 2022, China has been developing AI legislation and cracking down on deepfakes and generative AI through national-level legislation. While China's Algorithmic Registry includes security assessments for registered algorithms, in practice it appears to be more focused on fostering a business-friendly AI market than upholding citizens' rights (Yang 2024). Meanwhile, in March 2024, Mexico's national government proposed an AI Law aimed at regulating the development, marketing and use of AI systems, particularly with a view to protecting human rights, ensuring the ethical and responsible use of data, and safeguarding data privacy. However, across these debates and legislative efforts, the focus largely centres on information extraction and reproduction – how generative AI exploits user data for often misleading or manipulative purposes – while overlooking the spatial and territorial effects of AI-driven urban governance.

In examining the significance and reach of AI within urban space, Cugurullo et al. (2023) argue that artificial intelligence thrives particularly in the city, where a critical mass of data is available to harvest and operationalise for pluriversal AI futures. Examining a range of AI-driven technologies – including autonomous vehicles, drones, robots, governance platforms and software algorithms – they suggest that despite the aspirational urban futures represented by AI, these technologies are situated in the simultaneous history of urban and technological developments. Muldoon et al. (2024) also note that any attempt to grant agency to AI in governance must recognise that the logics of automation and predictability within AI futures require an enormous array of digital labour that is often precarious and casualised – akin to the peripheral labour within the digitalising state (Datta and Muthama, Chapter 2). As a result, AI futures are plural, messy and relational. On this point, Söderström and Datta remind us that with any type of data technology, we need to be wary of universalist notions which assume that 'data deluge would unfold in the same way with the same consequences everywhere' (Söderström and Datta 2023, 4). Instead, each wave of new technologies produces and reconfigures informational peripheries, embedding them further within the contexts and histories from which these technologies emerge. Hence, there is a need to understand how these technologies both operate in and structure contemporary governance and capitalism in the Global North and South (Söderström and Datta 2023).

One way to consider how informational peripheries are reproduced with the introduction of AI technologies is to start from the understanding that the calculability and predictability of AI futures thrive on historical data concentrations, which are most likely to be found in urban centres. AI technologies will configure themselves and their contexts in specific ways that are derived from the extrapolation of pre-existing data, which is unlikely within informational peripheries. This is evident in several aspects of AI urbanism – driverless cars, drones and robots are solidly situated in city centres or high-traffic spaces where they can extract and extrapolate data for their own verifiability. From that perspective, AI futures reinforce pre-existing informational peripheries, embedding themselves within the social and structural baggage of cities while also deepening their reliance on data silos.

Another approach is to consider AI's impact on the informational peripheries that have been bypassed in the ubiquity of smart urban technologies and yet are fully connected to a global informational space through a range of AI software. In this book, these spaces are characterised by hyper-concentrated and hyper-dispersed infrastructures, such as in the case of the logistics corridor in Bhiwandi near Mumbai (Chapter 4) or the migrant waste labourers in Ahmedabad (Chapter 3). Similar spaces exist globally, in the migrant neighbourhoods of cities like London, Dubai and New York, where the invisibility of rough sleepers in state databases for welfare and housing coexists with the widespread use of generative AI to track and police their movement. Ultimately, the aspirational embrace of AI technologies across both the Global North and South underscores the boundless extractive nature of these technologies and suggests the potential for multiple, overlapping informational peripheries that transcend geographic and digital boundaries yet are rooted in urban histories. As Maalsen et al. (2023) argue, we will need strong feminist, queer and geographical research on the development and uptake of AI to understand how it intersects with current structural inequalities, how it is contested from the margins, and how it lands in space.

Future informational peripheries

This volume has presented the digitalisation of governance and territories as the production of multifaced 'informational peripheries' that disclose both the operational and extractive logics of the digitalising state and the possibilities for more democratic and localised uses of digital technologies within urban space. The chapters in this book have highlighted the diverse actors producing and sustaining the informational periphery as they cope with and manage the simultaneous onslaught of information absence and overload in a digital age, even when they are marginalised in the process.

But what is specific to informational peripheries that reconfigures urban theory in a digital age? In her analysis of the geopolitics of knowledge production, Connell argued:

> The role of the periphery is first to supply data, and later to apply knowledge in the form of technology and method. The role of the metropole, as well as producing data, is to collate and process data, producing theory (including methodology), and developing applications that are later exported to the periphery. (Connell 2017, 6)

In this concluding chapter, we have shown how the impact of informational peripheries on theorising the urban is both incremental and radical: incremental because it extends the reach of Southern theory into informational spaces in the North and the future, and radical because it questions what the Global South or the periphery is or indeed does as a conceptual space in urban theory when information in a digital age is both boundless and bounded through specific processes, pathways and networks.

Extending the insights of this book beyond these pages, we argue that understanding the future significance and impact of digitalisation on the city requires a deeper exploration of the trajectories leading to full, partial and fragmented AI futures across the Global North and South. The contingencies, dynamism, multiplicity and cumulative nature of information infrastructures are constantly producing new and different peripheries. The chapters in this book have shown how this happens in both small ways – for example through routines and habits during a pandemic – and on a large scale – such as through the reproduction of apartheid histories. Pulled together, we believe they provide a conceptual and empirical starting point for unpacking the processes of informational peripheralisation as new technologies are folded into urban space.

References

Amoore, L. 2018. Cloud geographies: Computing, data, sovereignty. *Progress in Human Geography* 42(1): 4–24. https://doi.org/10.1177/0309132516662147.

Barns, S. 2016. Mine your data: Open data, digital strategies and entrepreneurial governance by code. *Urban Geography* 37(4): 554–71. https://doi.org/10.1080/02723638.2016.1139876.

Batty, M. 2018. Artificial intelligence and smart cities. *Environment and Planning B: Urban analytics and city science* 45(1): 3–6. https://doi.org/10.1177/2399808317751169.

Benjamin, R. 2019. *Race after Technology: Abolitionist tools for the New Jim Code*. Cambridge: Polity Press.

Bhatia, A. and Bhabha, J. 2017. India's Aadhaar scheme and the promise of inclusive social protection. *Oxford Development Studies* 45(1): 64–79. https://doi.org/10.1080/13600818.2016.1263726.

Branch, J. 2017. Territorial conflict in the digital age: Mapping technologies and negotiation. *International Studies Quarterly* 61(3): 557–69. https://doi.org/10.1093/isq/sqx046.

Bridges, L. E. 2024. Competing digital capacities: Between state-led digital governance and local data center tradeoffs. *Information, Communication & Society* 27(1): 1906–23. https://doi.org/10.1080/1369118X.2024.2331765.

Brodie, P. 2023. Data infrastructure studies on an unequal planet. *Big Data & Society* 10(1). https://doi.org/10.1177/20539517231182402.

Calvo, P. 2020. The ethics of smart city (EoSC): Moral implications of hyperconnectivity, algorithmization and the datafication of urban digital society. *Ethics and Information Technology* 22(2): 141–9. https://doi.org/10.1007/s10676-019-09523-0.

Caranto Morford, A. and Ansloos, J. 2021. Indigenous sovereignty in digital territory: A qualitative study on land-based relations with #NativeTwitter. *AlterNative: An international journal of indigenous peoples* 17(2): 293–305. https://doi.org/10.1177/11771801211019097.

Connell, R. 2017. Southern theory and world universities. *Higher Education Research & Development* 36(1): 4–15. https://doi.org/10.1080/07294360.2017.1252311.

Couldry, N. and Mejias, U. A. 2019. *The Costs of Connection: How data is colonizing human life and appropriating it for capitalism*. Redwood City, CA: Stanford University Press.

Cugurullo, F., Caprotti, F., Cook, M., Karvonen, A., McGuirk, P. and Marvin, S. 2023. *Artificial Intelligence and the City: Urbanistic perspectives on AI*. Abingdon: Routledge.

Datta, A. 2018. The digital turn in postcolonial urbanism: Smart citizenship in the making of India's 100 smart cities. *Transactions of the Institute of British Geographers* 43(3): 405–19. https://doi.org/10.1111/tran.12225.

Datta, A. 2020. Self(ie)-governance: Technologies of intimate surveillance in India under COVID-19. *Dialogues in Human Geography* 10(2): 234–7. https://doi.org/10.1177/2043820620929797.

Datta, A. and Muthama, D. M. 2024. Sorting paper: Archival labour of digitising land records in Kenya. *The Geographical Journal* 190(4). https://doi.org/10.1111/geoj.12581.

Dattani, K. 2020. 'Governtrepreneurism' for good governance: The case of Aadhaar and the India Stack. *Area* 52(2): 411–19. https://doi.org/10.1111/area.12579.

de Sousa Santos, B. 2016. Epistemologies of the South and the future. *From the European South* 1: 17–29. Accessed February 2025. https://www.fesjournal.eu/wp-content/uploads/2016/09/3.2016-1.Santos.pdf.

Edwards, L., Martin, L. and Henderson, T. 2018. Employee surveillance: The road to surveillance is paved with good intentions. *SSRN Electronic Journal*. https://doi.org/10.2139/ssrn.3234382.

Eubanks, V. 2018. *Automating Inequality: How high-tech tools profile, police, and punish the poor*. New York: St Martin's Press.

Gururani, S. 2024. Paradoxes of the periphery. *Asian Geographer* 41(2): 143–50. https://doi.org/10.1080/10225706.2024.2340966.

Halvorsen, S. 2018. Cartographies of epistemic expropriation: Critical reflections on learning from the south. *Geoforum* 95: 11–20. https://doi.org/10.1016/j.geoforum.2018.06.018.

Herath, H. M. K. K. M. B. and Mittal, M. 2022. Adoption of artificial intelligence in smart cities: A comprehensive review. *International Journal of Information Management Data Insights* 2(1). https://doi.org/10.1016/j.jjimei.2022.100076.

Hoefsloot, F. I., Martínez, J. and Pfeffer, K. 2024. An emerging knowledge system for future water governance: Sowing water for Lima. *Territory, Politics, Governance* 12(6): 825–45. https://doi.org/10.1080/21622671.2021.2023365.

Kitchin, R. and Dodge, M. 2017. The (in)security of smart cities: Vulnerabilities, risks, mitigation, and prevention. *Journal of Urban Technology* 26(2): 47–65. https://doi.org/10.1080/10630732.2017.1408002.

Maalsen, S., Cinnamon, J. and Kinsley, S. 2023. Artificial intelligence, geography and society. *Digital Geography and Society* 4. https://doi.org/10.1016/j.diggeo.2023.100061.

Mattern, S. 2021. *A City is Not a Computer: Other urban intelligences*. Princeton, NJ: Princeton University Press.

Mello Rose, F., Thiel, J. and Grabher, G. 2022. Selective inclusion: Civil society involvement in the smart city ecology of Amsterdam. *European Urban and Regional Studies* 29(3): 369–82. https://doi.org/10.1177/09697764221092587.

Muldoon, J., Graham, M. and Cant, C. 2024. *Feeding the Machine: The hidden human labour powering AI*. Edinburgh: Canongate Books.

Robinson, J. 2011. The spaces of circulating knowledge: City strategies and global urban governmentality. In *Mobile Urbanism*, edited by E. McCann and K. Ward, 15–40. Minneapolis, MN: Minnesota University Press.

Robinson, J. 2022. Introduction: Generating concepts of 'the urban' through comparative practice. *Urban Studies* 59(8): 1,521–35. https://doi.org/10.1177/00420980221092561.

Robinson, J. and Parnell, S. 2011. Traveling theory: Embracing post-neoliberalism through southern cities. In *The New Blackwell Companion to the City*, edited by G. Bridge and S. Watson, 521–31. Chichester: John Wiley & Sons.

Sarkar, S. 2014. The unique identity (UID) project, biometrics and re-imagining governance in India. *Oxford Development Studies* 42(4): 516–33. https://doi.org/10.1080/13600818.2014.924493.

Sassen, S. 2014. *Expulsions: Brutality and complexity in the global economy*. Cambridge, MA: Harvard University Press.

Söderström, O. and Datta, A. 2023. *Data Power in Action: Urban data politics in times of crisis*. Bristol: Bristol University Press.

Stikker, M. and van Eeden, Q. 2021. Komst datacenters verloopt ondoorzichtig en ondemocratisch. *Het Financieele Dagblad*, 10 December. Accessed February 2025. https://fd.nl/opinie/1422605/komst-datacenters-verloopt-ondoorzichtig-en-ondemocratisch-ijl1aaTZWxJL.

Taylor, L. 2017. What is data justice? The case for connecting digital rights and freedoms globally. *Big Data & Society* 4(2). https://doi.org/10.1177/2053951717736335.

Tsing, A. L. 2005. *Friction: An ethnography of global connection*. Princeton, NJ: Princeton University Press.

Vonderau, A. 2019. Scaling the cloud: Making state and infrastructure in Sweden. *Ethnos* 84(4): 698–718. https://doi.org/10.1080/00141844.2018.1471513.

Woodman, D. and Cook, J. 2019. The new gendered labour of synchronisation: Temporal labour in the new world of work. *Journal of Sociology* 55(4): 762–77. https://doi.org/10.1177/1440783319879244.

Yang, Z. 2024. Four things to know about China's new AI rules in 2024. *MIT Technology Review*, 17 January. Accessed February 2025. https://www.technologyreview.com/2024/01/17/1086704/china-ai-regulation-changes-2024.

Zuboff, S. 2015. Big other: Surveillance capitalism and the prospects of an information civilization. *Journal of Information Technology* 30(1): 75–89. https://doi.org/10.1057/jit.2015.5.

Zuboff, S. 2023. *The Age of Surveillance Capitalism: The fight for a human future at the new frontier of power*. London: Profile Books.

14

Epilogue: of digital frontiers and peripheries

D. Asher Ghertner

The urban peripheries are transitional spaces, fractally marked by connections to and dislocations from the core cultural, economic, political and infrastructural nodes that define the urban experience and give meaning to the idea of city-ness. My current home in the coastal state of Goa, India, provides a simple picture of this fractal quality. On one side of the small, gated complex where I have been renting what locals call a 'settler villa' lie common, seasonally flooded lands known as *khazans*, where rice is grown during the monsoon. The building belonging to the village panchayat, India's constitutionally enshrined rural governance body, sits adjacent to this area. An industrial estate with the region's only global pharmaceutical plant towers over the village on a hill. To reach the estate by road, one must pass through *comunidade* land, held by a pre-colonial agrarian landowning institution over which neither the panchayat nor the state government has control. City-style apartment complexes have sprouted up along the roadside of this semi-forested area, with manicured front gardens and a café that could belong to any global metropolis. The largest 'city' in the district houses a mere forty thousand people, and squat village bungalows, fields and bespoke vacation homes for metropolitan exiles dot the landscape stretching to the capital 15 kilometres south. The city is not the centre here, as this urbanising state has attracted these new settlers for what journalist Mukul Kesavan (2018) has called its enduring rural cosmopolitanism.

Seated in the café, digital nomads (and sabbatical professors) look out peripherally from their laptops across the paddy fields, watching sari-clad women hand-weed. But the digital periphery undergoes complex folds here. Goa was the first Indian state to digitise its land records, but *comunidades*, which control around 70 per cent of village land in the state, represent blank spots on the cadastral map. They lie beyond the synoptic gaze of property portals and the state's online land registry. As real-estate pressures on ecologically sensitive land have driven dozens of court cases in this one village alone, the *comunidade* conducted drone surveys to create a data bank to defend customary land uses from the circling development sharks. Land here operates as customary institution, digital portfolio and ecological amenity. Step one way and land falls out of clock time and is structured by the rhythms of planting, weeding and harvesting; step the other way and it is unavoidably calculable on the terms of what Manuel Castells (2009, 34) calls 'timeless time' – the structural ephemerality characteristic of the 'network society', where digital technologies enable the practice of 'simultaneity without contiguity'.

Providing an invaluable conceptual addition to how we reckon with the fractal geographies of rural-urban zones under conditions of intensified digital mediation, *Informational Peripheries* aptly describes the space 'where the rhythms of global capital, information infrastructures and local land politics meet and propel the tensions and contradictions of urban transformation'. Across 12 theoretically innovative chapters, the volume argues that the informational periphery is a product of territorialisation practices that remake once remote geographies as networked absences or presences. Putting empirical meat on the bones of rising calls to consider digital geographies from the Global South (Shaw and Graham 2017), the book retunes attention to the contested territoriality of the urban fringes where so many insurgent political forces – both progressive and regressive, old and new – are playing out. It does so by showing how digital infrastructures create new forms of territorial authority that can both reinforce and weaken traditional power centres – be they land bureaucracies, political strongmen or agrarian associations.

The volume uses three spatial-analytical lenses to organise its chapters: *the production of the informational periphery*, which focuses on labour and bureaucratic practices of bringing urban peripheries into digital legibility; *the territorialisation of the periphery*, which analyses competing authorities over how digital knowledge is governed; and what we can think of as *the periphery from below*, with an emphasis on hacks and non-statist applications of, or experiences with, geographical information and its infrastructures. For this short epilogue, I want to explore how two geographical concepts – namely, the network and the frontier – traverse the book's chapters in parallel to ask how digitalisation affects not just knowledge and control of land, but the very social imaginary of what land is.

The network

While the spatial metaphor of the cloud may have superseded the network in the contemporary digital imaginary (Chown and Nascimento 2023), network thinking continues to pervade how we understand urbanisation processes. As Datta and Hoefsloot write in Chapter 1 of this volume: 'In the informational periphery, social exclusions are marked by both geographic and informational distance from the state; it includes subjects that are uncountable as well as territories that are unmappable – digitally, socially and geographically.' This formulation has clear resonances with Castells' (1996) influential account of the network society, which, he postulates, emerges via instantaneous flows of capital, information and symbols – a 'space of flows' around which time is rendered timeless. The space of flows is formed, in his account, by a continuous and expanding circuit of digital information, which in turn subsumes situated places without eliminating their singularity. As Castells puts it, 'dominant functions are organised in networks pertaining to a space of flows that links them up around the world, while fragmenting subordinate functions and people in the multiple spaces of places, made up of locales increasingly segregated and disconnected from each other' (Castells 1996, 476).

Datta (Chapter 6), in discussing Bhiwandi's informational integration into logistical networks in greater Mumbai, begins with a similar account of inequality as a networked division 'made through uneven and unequal access to networked connectivity, as concentrated nodes of informational space emerge within longstanding structural disadvantages'. Chung, Dai and Xu (Chapter 7), describing the transformation of Guiyang, China into 'a big data valley', similarly find that the informational periphery is marginalised by and for information flows. Politically removed from the core, this periphery is opened as a territory through which information flows, but about which information is not generated.

Yet, while observing entrenched digital inequalities, these chapters simultaneously resist reducing the informational periphery to the metaphor of the digital divide, or the dualism symptomatic of Castells' formulation of a disembodied global overriding local histories and geographies. Instead, they offer a relational notion of informational space, attending to how digital integration is the product of explicitly spatialised, power-laden processes that operate at multiple levels, including through local power brokers, bureaucrats and strongmen. Informational peripheries thus do not emerge tabula rasa via technological intervention. This is not the progressive colonisation of analogue time by new techno-scapes, to evoke the hyperbolic language of early globalisation debates (see Appadurai 1996). They rather form as 'stacks' – platforms of exchange built upon colonial-era, paper-based land revenue systems and agrarian economic geographies. As Datta (Chapter 6) puts it in relation to Bhiwandi's warehouse landscapes, 'global logistics settles in Bhiwandi as warehouses, and the local networks of caste-ist, ethnic and religious connections to land'. Throughout the book, we learn how the political economy of information, and hence territorial command over land and its relations, remains embedded in vernacular networks, with a particular grammatological rooting in linguistic and inscriptive conventions that require specialist land brokers-cum-translators and that rest on categorical openness ill-suited to digital simplification (see Ghertner 2024). Property, even in its ostensibly flattened digital form, continues to be underpinned by unruly categories, such as in Cape Town, where algorithmic property transactions are stacked upon colonial-era racial classifications that facilitate but also haunt business as usual, always risking eruptions of protest or rage against the machine (Migozzi, Chapter 8; von Schnitzler 2017).

In the case of the digital land record system known as 'Visor Urbano' in Guadalajara, Mexico (Demerutis Arenas, Datta and Flores, Chapter 4), we learn that common-property *ejido* and indigenous community lands are historically produced rural initiatives that fall outside the jurisdiction of urban municipalities. As in the *comunidade* land in Goa I mentioned above, when these areas are mapped, their uses are not incorporated into urban municipal databases. While Visor Urbano's exclusion of historic tenure and land-use information could be framed as a stubborn obstacle posed in a Castellsian space of places – an unintegrated and bypassed outside – Demerutis Arenas, Datta and Flores instead show how land operates as a political technology that 'seeps into the ontologies of digital information systems'. Visor Urbano's legacy as an administrative technology aimed at producing fungible property across territory inadvertently sets *land* apart from *property*, the former conditioned by residual *data* that cannot become transactional *information*. Here, the informational periphery is not an exclusion from the network but its negative condition – a frontier that ushers in a sovereign urge to conquer and control, but which necessarily reproduces competing authorities. The promise of digital simplification always frays and splinters. This is what Cowan (Chapter 9) calls 'peripheral property', or lands 'excised from or oppositional to the mainstream, computable grammars of urban real estate'. The periphery becomes the frontier when it is problematised as a conquerable space, a version of what Li (2014) calls 'land's end', where the promises of cadastral calculability meet the enduring tangle of agrarian life and ecology and where colonial technologies – be they the ethnographic surveys of old or modern drone-generated aerial imagery – promise a technological fix for political intractability.

The frontier

Frederick Jackson Turner's (1920) classic frontier thesis conceives of the frontier through the conditions of geographical isolation, plentiful land, untapped resources, and gradual settlement by commercially driven pioneers and state authorities. However, Turner's account minimises the role of the accumulation process, including its associated violence and disorder, and ignores how frontier expansion is tied to the extension of state power. In contrast, Michael Watts (2017) argues that frontier regions involve ongoing struggles over what constitutes legitimate authority and who has the power to establish it, highlighting the contentious nature of rule in these spaces of resource struggle.

Frontiers, in Watts's formulation, can be thought of as aspects of what Carl Schmitt called *Landnahme* (see Graf 2022), or the land-appropriating state. This European territorial force was thrust into history by the Western colonial confrontation with 'other' states beyond European rationality/nomos in the late nineteenth century (including the US). It emerged, then, as a response to the 'spacelessness' of international law in the wake of the European 'age of discovery' by applying the rule of exception to the undefined territorial outside (see Pincince 2014). Frontiers and state-backed violence, in this formulation, are inherently intertwined, and become central to international political-economic shifts or deterritorialised geopolitical arrangements – be they moments of a global land rush, as in the mid-2000s commodity boom; technological leap-frogging, as in the pervasiveness of real-time, cloud-based, remote monitoring services; or shifts in geo-strategic relations, as varied as the global conflagrations associated with decolonisation or 'global China's' pursuit of resource control.

In what ways is the informational periphery, as framed in the present volume, also a digital frontier, a space of claims making that operates through exceptional powers to suspend political confrontation and supersede legal precedent in the name of a higher order – a promised digital future? A frontier framework is indeed implicit throughout the volume, reflected in Chapter 1's framing of the informational periphery as a 'paradoxical' place that is 'at once marked by invisibility, exclusion and marginalisation while at the same time being targeted for incorporation into the territory of the metropolitan city'. All of the chapters, in one way or another, analyse the territorialisation of the informational periphery through extractive processes – be they related to data, environmental resources, labour or square metres of land. As such, it is here – in the attempt to overcome opacity, to render 'low resolution' peripheries into quantifiable property, and to normalise propertied forms of citizenship – that digitalisation must be understood as a frontier technology. As Cowan (Chapter 9) asserts, such efforts pursue not the more efficient management of property assets, but rather the removal of tenure diversity and the creation of new property assets. Highlighting this as frontier work is crucial, for it lays bare the efficiency claims through which property technologies and land digitalisation programmes are sold, returning focus to the territorialised violence inherent in frontier processes (Ghertner 2020). It simultaneously reminds us that, unlike in most Global North contexts where property technologies are market-driven and oriented toward the acceleration or financialisation of already marketised relations, in the Global South, digital infrastructures are largely extensions of state schemes to break down customary barriers to private property. They are state-territorialising actions, albeit couched in the developmental goals of market-based efficiency and technological inclusion. If the periphery is understood as a space of lack and deprivation, then the frontier holds the potential for conquer and extraction. The periphery is kept at a distance, while the frontier draws territories nearer. State-driven digitalisation schemes sit at this intersection.

Yet the focus on digitalisation is not just about new geographies of capture/banishment conditioned by informational intensification. Understanding such processes through a frontier framework also draws attention to new and contested subjectivities. I'm thinking here of how certain historical subjects, especially those 'residual' categories of colonial and feudal political-economic orders – including agrarian landed classes, customary landholders, village strongmen and urban villagers – come to be imagined as potential urban citizens, subject to their submission to urban property forms. These are improvable subjects, a primary target of all frontier schemes. But I'm also thinking of non-improvable subjects, including the landless precariat, migrants, and other potentially wasted lives-cum-populations-of-the-banished defined by their alterity vis-à-vis the rules of formal ownership. In the case of the former group, we see the execution of forms of what Safransky (2020) terms algorithmic violence in the automated property valuation and credit scoring in the Cape Town that Migozzi (Chapter 8) describes. There, Black and

Coloured townships are programmed as low-value territories based on their historic and ongoing subjection to the police state. In India, where property technologies operate via existing biometric identification and citizenship technologies, remaking peripheral subjects on the terms of cadastral normativity and plot-linked homeownership allows the state to curtail rights and entitlements flexibly. Cowan (Chapter 9) frames this transposition as bringing citizenship into degrees of alignment with received property data categories. In Mexico City, even in contexts where the goals of repair emerged through state-society exchange, Boudreau (Chapter 11) finds 'street epistemologies' written out of core logics of outreach, causing legitimate gestures toward empathy to be overwritten by the cold calculus of community accounting. In Lima, Hoefsloot (Chapter 12) finds that infrastructural injustices around water are reproduced via digital exclusions, rendering those with less speculative property values peripheral to state water schemes.

For the unpropertied precariat subjected to new marketised land relations and propertied norms of citizen conduct, the potential to render whole segments of the population ungeographic, or placeless in the city, is still higher. Wittmer's focus (Chapter 3) on the subsumption of waste labour in Ahmedabad, for example, shows how waste management practices are increasingly becoming 'datafied', or quantified through digital technologies, to extract maximum value from discarded materials and wasted bodies. The same 'informational invisibility' that submits workers to the logistical optimisation of recycling companies that Wittmer discusses is evident among the migrants in urban villages outside of Delhi, where Goldstein (Chapter 10) finds that informational citizenship enables migrants' portability at the expense of their ability to be counted as locals. In other words, informational peripheries not only constitute frontier territories – or spaces of territorialised violence and competing authority – but frontier subjects, defined by their lack of emplacement, which itself becomes a resource (Bathla 2022), a tool to be manipulated both to extract surplus value and to pin the costs of social reproduction on other geographies (the village 'back home', the state next door, the translocal household (Gidwani and Ramamurthy 2018)).

Differing from a 'space of places', or a networked exclusion, the chapters in this volume show a different kind of frontier subject occupying the residual spaces of the informational periphery, dwelling in a virtual potentiality, 'the surrounds' (Simone 2022), that functions as a negative space defined by an unlocatability, a shifting terrain of relational possibility that slips away just as it seems amenable to capture. The countable, biometric subject demurs just as the camera focuses; a chalk line is kicked away as a drone flies above; an indistinguishable set of shanties defy the parsing gaze of the satellite imagery; the surveyor's list of discrete households gains an asterisk after an internal wall is discovered marking two houses where only one was recorded. A map becomes a picture upon which drawings are traced in a bureaucrat's folder, marking the houses earmarked for resettlement, and always remaining open to amendment. A waste worker tosses their NGO-issued ID card yet continues to recycle the streets of their neighbourhood. In these peripheries, on this frontier, where politics is meant to be coded out of the equation, analogue lifeworlds reproduce turbulent possibilities. From my café in Goa, the monsoon rains knock out the internet, and the harvesting goes on. Localised information resists complete abstraction, yes, but, as significantly, digital abstractions gain traction through their uptake in sites of use.

References

Appadurai, A. 1996. *Modernity at Large: Cultural dimensions of globalization*. Minneapolis, MN: University of Minnesota Press.

Bathla, N. 2022. Planned illegality, permanent temporariness, and strategic philanthropy: Tenement towns under extended urbanisation of postmetropolitan Delhi. *Housing Studies* 37(6): 868–88. https://doi.org/10.1080/02673037.2021.1992359.

Castells, M. 1996. *The Rise of the Network Society*. Oxford: Blackwell.

Castells, M. 2009. *Communication Power*. Oxford: Oxford University Press.

Chown, E. and Nascimento, F. 2023. *Meaningful Technologies: How digital metaphors change the way we think and live*. Mountain View, CA: Lever Press.

Ghertner, D. A. 2020. Lively lands: The spatial reproduction squeeze and the failure of the urban imaginary. *International Journal of Urban and Regional Research* 44(4): 561–81. https://doi.org/10.1111/1468-2427.12926.

Ghertner, D. A. 2024. Scripts, scribes and scribbles: Notes on drafting the South Asian city. *City* 28(1–2): 113–20. https://doi.org/10.1080/13604813.2024.2320005.

Gidwani, V. and Ramamurthy, P. 2018. Agrarian questions of labor in urban India: Middle migrants, translocal householding and the intersectional politics of social reproduction. *Journal of Peasant Studies* 45(5–6): 994–1017. https://doi.org/10.1080/03066150.2018.1503172.

Graf, S. 2022. Carl Schmitt reads Hannah Arendt's *Eichmann in Jerusalem*: Archival perspectives on convergences and divergences. *American Journal of Political Science 66(4): 918–31*. https://doi.org/10.1111/ajps.12682.

Kesavan, M. 2018. Desis in Goa. *Telegraph India*, 15 September. Accessed February 2025. https://www.telegraphindia.com/opinion/desis-in-goa/cid/1668760.

Li, T. M. 2014. *Land's End: Capitalist relations on an indigenous frontier*. Durham, NC: Duke University Press.

Pincince, J. 2014. De-centering Carl Schmitt: The colonial state of exception and the criminalization of the political in British India, 1905–1920. *Política común* 5. https://doi.org/10.3998/pc.12322227.0005.006.

Safransky, S. 2020. Geographies of algorithmic violence: Redlining the smart city. *International Journal of Urban and Regional Research* 44(2): 200–18. https://doi.org/10.1111/1468-2427.12833

Shaw, J. and Graham, M. 2017. An informational right to the city? Code, content, control, and the urbanization of information. *Antipode* 49(4): 907–27. https://doi.org/10.1111/anti.12312.

Simone, A. 2022. *The Surrounds: Urban life within and beyond capture*. Durham, NC: Duke University Press.

Turner, F. J. 1920. *The Frontier in American History*. New York: Holt and Company.
von Schnitzler, A. 2017. *Democracy's Infrastructure: Techno-politics and protest after apartheid*. Princeton, NJ: Princeton University Press.
Watts, M. J. 2017. Frontiers: Authority, precarity and insurgency at the edge of the state. *L'Espace politique* 32. https://doi.org/10.4000/espacepolitique.4336.

Index